THE NIGHT IT RAINED GUNS

A senior assistant editor at *The Times of India*, **Chandan Nandy** has been an investigative reporter for *The Telegraph* and the *Hindustan Times* in a journalistic career spanning nearly twenty years. He has also covered India's intelligence agencies, the home ministry, internal security and defence. He graduated from Brandeis University, Massachusetts, with a masters in Coexistence and Conflict Resolution in 2006 and also studied at what was formerly Presidency College in Kolkata.

THE NIGHT IT RAINED GUNS

UNRAVELLING THE PURULIA ARMS DROP CONSPIRACY

CHANDAN NANDY

FOREWORD BY
VIR SANGHVI

RUPA

Published by
Rupa Publications India Pvt. Ltd 2014
7/16, Ansari Road, Daryaganj
New Delhi 110002

Sales centres:
Allahabad Bengaluru Chennai
Hyderabad Jaipur Kathmandu
Kolkata Mumbai

Copyright © Chandan Nandy 2014
Photographs courtesy author archives

ISBN: 978-81-291-3472-1

First impression 2014

10 9 8 7 6 5 4 3 2 1

The moral right of the author has been asserted.

Typeset in Palatino Linotype by SÜRYA, New Delhi

Printed at Replika Press Pvt. Ltd, India

Dedicated to the memory of my father,
Bibhuti Bhushan Nandy (1940–2008).

Wish you were around.

CONTENTS

LIST OF ABBREVIATIONS

AAI	Airports Authority of India
ACIU	Analytical Criminal Intelligence Unit (Interpol)
ADC	Air Defence Clearance
ARC	Aviation Research Centre
ASI	Assistant Sub-Inspector
ATF	Bureau of Alcohol, Tobacco, Firearms and Explosives (United States)
AVM	Air Vice-Marshal
BJP	Bharatiya Janata Party
BNP	Bangladesh Nationalist Party
BTI	Border Technology and Innovations Ltd
CBI	Central Bureau of Investigation (India)
CCA	Controller of Certifying Authority (Bangladesh)
CEO	Chief Executive Officer
CIA	Central Intelligence Agency (United States)
CPI (M)	Communist Party of India (Marxist)
CVR	Cockpit Voice Recorder
DEA	Drug Enforcement Administration (United States)
DESO	Defence Export Service Organisation (United Kingdom)
DGCA	Directorate General of Civil Aviation (India)
DGDP	Directorate General Defence Purchase (Bangladesh)
DGFI	Directorate General of Forces Intelligence (Bangladesh)
DO (letter)	Demi-Official
ECHR	European Commission on Human Rights
EUC	End-User Certificate
FARC	Fuerzas Armadas Revolucionarias de Colombia
FDR	Flight Data Recorder
FIC	Flight Information Centre (Chennai, India)
FIR	First Information Report
FOB	Free On Board
FRU	Force Research Unit (British Army)
GPS	Global Positioning System
HSBC	Hong Kong and Shanghai Banking Corporation
IAF	Indian Air Force
IB	Intelligence Bureau (India)
IO	Investigating Officer
IPC	Indian Penal Code

IRA	Irish Republican Army
ISI	Inter-Services Intelligence (Pakistan)
KGB	Komitet Gosudarstvennoy Bezopasnosti (Russia)
KIA	Kachin Independence Army (Myanmar)
LTTE	Liberation Tigers of Tamil Eelam (Sri Lanka)
MCC	Maoist Coordination Committee (India)
MEA	Ministry of External Affairs (India)
MHA	Ministry of Home Affairs (India)
MLU	Military Liaison Unit (Indian Air Force)
MoD	Ministry of Defence (India)
MP	Member of Parliament
NATO	North Atlantic Treaty Organisation (Belgium)
NCB	National Centre Bureau (Interpol)
NGO	Non-Governmental Organization
NIA	National Investigation Agency (India)
NSCN	National Socialist Council of Nagaland (India)
OICD	Organised and International Crime Directorate (UK Home Office)
PAF	Pakistan Air Force
PET	Politiets Efterretningstjeneste (Denmark)
PIRA	Provisional Irish Republican Army
R&AW	Research and Analysis Wing (India)
RAX	Restricted Area Exchange
SALW	Small Arms and Light Weapons
SAS	Scandinavian Airline System
SIS	Secret Intelligence Service (or MI6, United Kingdom)
SLO	Security Liaison Office (MI5, United Kingdom)
SP	Superintendent of Police
SPLA	Sudan People's Liberation Army
UAE	United Arab Emirates
UK	United Kingdom
ULFA	United Liberation Front of Assam (India)
UN	United Nations Organization
UNITA	União Nacional para a Independência Total de Angola
UO (note)	Unofficial note
US	United States of America
USSR	Union of Soviet Socialist Republics
VHGR	Velocity Height Gravity Recorder
VSS	Voluntary Security Service (India)

FOREWORD

No Indian can take much pride in the events surrounding the Purulia arms drop of December 1995. The Indian government was warned by the British security services that an aircraft would deliver arms to the region. India's external intelligence agency, the Research and Analysis Wing (R&AW), processed the information and sent out warnings to various domestic officials and agencies.

But the information was ignored and no action was taken. The arms drop went ahead anyway. Then, when the aircraft made its return flight over India some days later, authorities failed at first to intercept it, even though it lacked the proper clearances. When the aircraft was finally forced to land at Mumbai airport, there was a terrifying absence of coordination and the principal organizer of the operation, a man who called himself Kim Peter Davy, simply walked out of the airport and escaped.

The Indian authorities then arrested the plane's Latvian crew—who knew little about the entire operation—and turned on Peter Bleach, a British arms dealer and probable part-time asset of the British Secret Services who had first passed on the tip-off about the airdrop.

In the manner of all Indian police forces and investigative agencies, having failed to act on perfectly good information provided by Bleach, the British Security Services and R&AW, the cops constructed a bogus case on the basis of which Bleach and the Latvians received long jail sentences.

The obvious injustice of this turn of events led to international protests and, eventually, first the Latvians and then Bleach won their freedom after their respective governments intervened.

Indians will not be surprised to learn that despite this sad and

shameful series of failures and foul-ups, not one official lost his job
and not one agency was held accountable. Instead, the bureaucratic
buck-passing got so nauseating that most of us just gave up on the
case.

Then, over a decade after the arms drop, Davy reappeared. This
time he used his real name: Niels Christian Nielsen. Far from being
ashamed of his status as an international fugitive, he strutted around
his native Denmark, defying the Indian authorities to come and get
him.

In the tragic and predictably inevitable manner that is the hallmark
of this story, India failed to get him extradited from Denmark. Danish
judges simply turned down Indian extradition requests.

But Davy-Nielsen was not done. He took to appearing in the
Indian media and spinning fanciful conspiracy theories. Among his
co-conspirators, he claimed, was an Indian politician who had helped
get the radars switched off so that the arms drop could go ahead. The
entire operation had been blessed (or even engineered) by New
Delhi a means of creating an armed insurrection, which would be
used as an excuse to dismiss the Left Front government in West
Bengal. There was nothing wrong with working against Joyti Basu's
regime anyway: women and children had suffered because of the
CPI(M). And so on.

Because we are still so fascinated by the Purulia mystery, Davy-
Nielsen probably got more publicity than he deserved. Yes, his stories
were incredible. But then, what was the truth? The Indian government
had done such a bad job of investigating the case that we were no
longer sure what to believe.

In this book, Chandan Nandy, who first followed the case as a
reporter for *The Telegraph* and then at the *Hindustan Times* (where I was
his editor), attempts to tell us what really happened. His account is
packed with facts and full of telling details. He exposes the spectacular
incompetence of the Indian authorities and tries to solve some of the
mysteries surrounding the case.

Was Davy-Nielsen an Ananda Margi? The answer is that he
certainly had some Margi connections and his lifestyle (he was a
vegetarian, for instance) suggested Margi beliefs. But it does not

necessary follow, as Nandy demonstrates, that the arms were meant for the Ananda Marga. Certainly, the nature and quantity of the weapons that were airdropped suggest that the Ananda Marga would have had little use for them.

So who were they really meant for? Could Burma's Kachin rebels have been the likely recipients? And was there a Western secret service involved in the operation on a hands-off, full-deniability basis? Could this involvement be the reason why Davy-Nielsen appears to have escaped any retribution?

We still don't have the answers to these questions, but thanks to this book, we now have the facts we need to make an informed assessment.

VIR SANGHVI
24 January 2013

INTRODUCTION

On 18 December 1995, Calcutta's (now Kolkata) young and bold English newspaper, *The Telegraph*, where I then worked as a senior reporter, would have missed the Purulia arms drop story if it were not for an alert sub-editor. He had walked up the stairs to the third floor to check the proof versions of the pages when his eyes fell on the day's issue of *Anandabazar Patrika*, the flagship Bengali publication of the ABP Group. Spread across eight columns on page one, the headline, in seventy-point bold font, screamed 'Mystery weapons in Purulia'. Below the headline was a large photograph of lethal weapons strewn across a field.

It was an amazing and bewildering case whose details slowly emerged: an ageing Russian Antonov AN 26 transport plane, carrying over a four-ton cache of lethal arms and ammunition, took off from Karachi on the evening of 17 December 1995. Its destination was Dhaka. On board the Antonov-26 were eight men—a Danish gun-runner, Kim Peter Davy, who was subsequently found to be the operation's mastermind; a British arms dealer and part-time source of Britain's MI6, Peter Bleach; a Singaporean of Indian descent, Deepak Manikan; and five Russian-speaking Latvians, Alexander Klichine alias Sasha, Igor Moskvitin, Oleg Gaidach, Evgueni Antimenko and Igor Timmerman. The plane refuelled at Varanasi's Babatpur Airport as the final touches were given to rig three parachutes to the deadly cargo. After taking off from Varanasi, the plane made a course deviation over Gaya, and when it was right over Purulia, a backward, nondescript district in West Bengal, it dropped down dangerously low. Three weapons-laden wooden pallets, fastened to the parachutes, were flung out of its belly into the night sky, dropped over the villages around Jhalda and Joypur police stations, very close to the global headquarters of the Ananda Marga, called Ananda Nagar. Mission accomplished, the aircraft returned to its original flight corridor, landed at Calcutta's Dum Dum airport, refuelled and flew off to Phuket in Thailand.

That night, the sub-editor ran back to *The Telegraph's* newsroom and shared his discovery conspiratorially, yet excitedly, with the bureau chief, Soumitra Banerjee, and chief reporter, Diptosh Majumdar. There was a huddle. A quick decision was taken: since it was already 11 p.m., it was too late to discard other lead stories to make room for *the* big story. That would have entailed redesigning the entire page. A delay in putting the edition to bed on time would hurt circulation the next morning. So a small, single-column story was quickly put together and slipped into the top half of the page before the edition was released.

There was something about the event that pulled me towards it when I read the arms drop story that was carried by every major Calcutta newspaper. I could sense there was more to the arms drop than met the eye. It was an enigma wrapped in mystery. It would be a real potboiler; the slow unveiling of a secret. There were inquiries to be made, a story to be told. That night, I swore to get to the bottom of the case. Within a few days I was to move to Jamshedpur to take charge of the bureau there, but that did not matter. I convinced Bobby, as the bureau chief was (and still is) fondly called by one and all in the noisy newsroom, to let me go after the story. He agreed.

In the next few days, I made preparations to shift to Jamshedpur. My mind, though, remained stuck on the mystery weapons. Two days went by, but none of the Calcutta newspapers, English or vernacular, were able to make any headway. Every reporter worth his or her salt threw shots in the dark. Some newspapers claimed, after reading the stencilled writings on the boxes containing the weapons— COMMANDANT CAD RAJENDRAPUR CANTT., BANGLADESH—that the weapons were destined for India's restive eastern neighbour. I wasn't quite convinced by all the theories that were floating about in the initial days after the arms were discovered by people in the quaint adivasi (indigenous people) villages around the area, under Jhalda and Joypur police stations, not far from the district headquarters. After quickly taking charge at *The Telegraph's* Jamshedpur bureau, I was back in Calcutta, calling up my friends in the Research and Analysis Wing (R&AW), India's external intelligence agency. I was pretty sure the weapons had originated somewhere abroad and the R&AW would have some, if not all, of the answers to the basic questions that I was confronted with at the time.

On 23 December 1995, Indian security officials reported the arrest of a British arms dealer, Peter Bleach, and five Latvian crew members of the Russian Antonov-26 transport aircraft from which the weapons were dropped. Nothing more was known. The authorities had taken a decision to shut up as the multi-agency interrogation of the arrested foreigners began in Mumbai.

It took me a few more days to get the first story break. On New Year's Eve, 1995, *The Telegraph* published the first[1] in a series of my stories on the Purulia arms drop, spanning over three years. This story provided details of the overtures made to Peter Bleach by two Danish nationals expressing their intent to purchase a large quantity of small arms and light weapons (SALWs); the name of a New Zealander, Kim Palgrave Davy, who was the sole buyer of the arms; the actual purchase of the weapons and the AN-26; and the names of the places—Copenhagen and Bangkok—where the conspirators met to firm up their plans. It also identified Burgas in Bulgaria as the location where the consignment was loaded onto the AN-26 before it flew to Karachi (and Varanasi en route) to deliver the weapons. The second investigative story appeared on 1 January 1996, with the headline 'Davy Top Gold Smuggler: CBI'.[2] Senior Central Bureau of Investigation (CBI) officers, including the then additional director, Upendra Nath Biswas, and joint director, Jyoti Krishna Dutt, who supervised the case at the time, were furious with the leaks. A few days later, a colleague and friend, Meher Murshed, joined me in the Purulia investigations. We teamed up and had a great time, breaking several more stories. There was no competition.

I confess today that the details of the first two stories—and later, others—did not originate from any source in the CBI, which, by 27 December 1995, had been entrusted with investigating the case, the first such in the history of independent India. That sleight of print was employed to disguise my real sources. Despite the sensitive details that Bleach shared with the authorities, the CBI's top management had no clear idea about how to proceed with the mass of information.

As a former R&AW chief later told me during a long conversation, 'When we first received the information from our British sources, it all sounded like a gravy plot.' Irrespective of how fantastic the plot sounded, the government and the domestic security agencies never mobilized in response to the R&AW's intelligence. They neither had the direction nor a plan they could institute and put into action to thwart the gunrunners. The airspace was not observed, communications systems and electronic surveillance were not utilized, intelligence resources were not activated and state and local law enforcement agencies were not sensitized—leave alone marshalled—to act on the information that the R&AW had supplied. Consequently, Nielsen and his handlers exploited the deep institutional cleavages and failings within the government and security agencies to their advantage. The obvious counter-factual question is whether enhanced vigilance would have led to an opportunity to disrupt the plot or even interdict the AN-26 when it was in Indian airspace, as well as at Varanasi airport. It was almost inconceivable that Indian airspace could be breached by a foreign aircraft to drop weapons with the aid of parachutes. India's national security had been violated with impunity and its putative protectors had messed up big time. To rub salt on their wounds, the kingpin and mastermind of the arms drop operation, Kim Davy aka Niels Christian Nielsen, had slipped through their fingers when he escaped under mysterious circumstances from Mumbai's Sahar International Airport the night the AN-26 aircraft was force-landed, after it was identified on 21 December 1995 by the Indian Air Force as the plane that might have dropped the weapons over Purulia.

Yet, here was a British arms dealer, Peter Bleach, who claimed he had informed British intelligence through every step of the conspiracy to procure the weapons and the aircraft, besides giving information on the kingpin. Bleach sang like a canary in the belief that his cooperation with the Indian authorities would ensure he went free, without charges being pressed against him. Of course, as the case panned out, Bleach found himself implicated, indicted, tried and then imprisoned for life before he was set free in March 2004 by a presidential decree of clemency, following a combination of sugar-coated requests and tacit pressures by the Labour government of Tony Blair. Bleach's cooperation and the information he supplied to the clueless investigators in the initial days of the probe, which was to take on transcontinental proportions in the months to come, was seminal. He had, undeniably, made some crucial contributions to the case and almost set the entire story out for the CBI's Special Crime Branch, whose job was to dig for clues and evidence which could then be strategically and tactically used to drive and sustain the prosecution stand whenever the case came up for trial in Calcutta. Unfortunately, the investigation was far from being airtight and satisfactory. The CBI was in a blind alley. The evidence collected, especially against the Ananda Marga cult[3], was spotty, and some of it did not stand the test of legal scrutiny. Other evidence unearthed to link four Ananda Marga monks who had actually collaborated with Kim Davy, also a member of the cult group, was wholly insufficient. The monks continue to remain fugitives from the law. The senior-most among them, Acharya Tadbhavananda, the then secretary general of the shadowy Ananda Marga unit called Proutist Universal, which advocated a strange mix of weird and muddled abstractions like 'neo-humanism' and 'economic democracy', was arrested in April 2004 in New Delhi. He fell prey to cancer seven months later, leaving unresolved the issue of his precise links with Nielsen—much of whose life was spent in the dark interstices of deception and suspected covert intelligence activities—and his likely knowledge about the end-users. I am in no doubt whatsoever that Nielsen had used the four Ananda Marga monks who were party to the larger arms drop conspiracy. But I seriously wonder whether the Ananda Marga as an organization, regardless of Nielsen's claims that he wanted to arm his cult brethren to defend themselves from the alleged depredations of West Bengal's communists, was the intended recipient of the weapons.

Nineteen years is a long time in the history of a criminal case which had all the elements of mystery, intrigue, drama, triumphs, frustrations, controversies, mires and abject failures. After the first five years of incessant and energetic pursuit of the case by the CBI, the investigation slowed down and then languished before it died; I dare say, it was allowed to die. After the departure of the ebullient CBI director, P.C. Sharma,[4] none of the succeeding chiefs had the appetite to plan and strategize newer operations to go after or ensnare the absconding suspects. The years from 2001 to 2011 were the wasted years. (In April 2011, though, Nielsen was arrested for a day in Copenhagen and then released on bail the very next day by the Danish police, which for years had protected him and stubbornly refused to execute an Interpol Red Corner Notice against him.) The global pursuit for evidence to unearth and identify the men or the organization behind the arms drop and the real end-users of the weapons could have continued with gusto. Inexplicably, it did not. It was a decade of absolute neglect and of utter drift. Files gathered dust, senior and middling officers with little or no imagination and bereft of a sense of history moved in and out of the CBI's Special Crime Branch, and investigators lost interest. They turned to other pressing cases and priorities. A few, who had not even tried to read the files related to past investigations, busied themselves with very narrow focus areas of the case as and when they arose—for instance, Nielsen's extradition proceedings in 2010 and 2011. The case did not resonate with some of the middle-level supervising officers who I tried to engage as I began writing the first draft of the book. Some shrugged and returned blank looks, some felt distinctly uncomfortable discussing controversial facts and issues, others hid behind the veil of ignorance and yet others fobbed me off.

The years between 2001 and 2011 were also reflective of institutional lag and indecisiveness as the security establishment, including the R&AW and the Intelligence Bureau (IB), the guardians of Indian national security, lost interest too, turning their attention to the looming threat of Islamist fundamentalism and terrorism that no longer just lurked on India's frontiers but had penetrated deep inside the country. The Sri Lankan civil war had reignited in the mid-1990s and the Maoist insurgency had erupted in Nepal. The intelligence failures of the 1999 Kargil war and the hijack of an Indian Airlines plane from Kathmandu the same year eclipsed past lapses, forcing the government of the time to take steps to reform the national security apparatus. At the same time, the security agencies' energies were sapped by internecine turf battles and, in the case of the R&AW, scandals of defection and other improprieties. Meanwhile, a few new security arrangements were established over a period of time so that efforts of intelligence agencies on battling terrorism could be analysed and coordinated effectively. But Purulia, arguably a complex and tough, and therefore a deeply fascinating case, was a forgotten chapter in the bleak and desultory history of India's national security. A combination of individual lack of dedication and institutional fatalism has prevented us from knowing the entire truth.

Part of the problem why the CBI's Purulia investigation came apart at the seams and individual officers lost steam after five years of admittedly commendable work was the agency itself. According to its charter, the CBI was formed mainly as an anti-corruption investigating body of the central government. Gradually, the scope of its mandate was widened to include investigating economic offences and important conventional crimes such as murders, abductions and terrorism on a 'selective basis'.[5] Indeed, the CBI pushed to expand its role in criminal cases with international linkages and limited aspects of terrorism, including home-grown insurgencies. Over time, as cases with political ramifications began to be entrusted to the CBI, its avowed claims of being an 'impartial' and 'competent' agency wore thin and it increasingly began to be viewed as a handmaiden of the government in power at the centre, which could suitably employ it to arm-twist troublesome political rivals within and the opposition outside. The centre's counter-terrorism policy was dispersed, in the sense that while it was the job of the IB and the R&AW to collect and disseminate intelligence on a case-by-case basis to its principal consumers, which included the Prime Minister's Office, the Cabinet Secretariat, the Home Ministry and the CBI, the investigating agency probed cases based on the information supplied to it and was expected to prosecute offenders. The intelligence agencies shared information on the established principal of 'need-to-know'. In most terrorism-related investigations, however, the CBI made requests to the R&AW and the IB to verify information it received from its own sources and produce intelligence in respect of specific cases. The IB continued to play the lead role in counter-terrorism and counter-espionage.

But the Purulia arms drop case was peculiar in more senses than one. Although intelligence that an AN-26 aircraft would be used by foreign mercenaries to drop weapons in the region of Dhanbad and Purulia had been available, the IB had failed in performing its duty to take adequate precautionary steps to prevent an incursion of the aircraft into Indian airspace, to send timely alerts and urgent advisories to the governments of West Bengal and Bihar and to thwart the escape of Niels Christian Nielsen from the Mumbai airport. In the backdrop of such security disasters, the CBI's Special Crime Branch tried to chart its own course on intelligence gathering, although it did rely a great deal on the R&AW, which was believed to have the resources to collect information from other countries. However, in the Purulia case, one crucial question remained unresolved: was it to be treated as a national security matter or as a law enforcement issue—a crime with international dimensions—that the CBI was capable of taking on? In 1995, as far as the political establishment— the Congress regime under the prime ministership of P.V. Narasimha Rao—was concerned, there was no conception of an overarching federal agency exclusively geared toward counter-terrorism. That would happen years later with the establishment of the National Investigation Agency (NIA) in December 2008. The Purulia case, inactive by then, continued to remain under the CBI's belt. I am not sure whether the country's national security bureaucrats ever discussed assigning it to the NIA. The political leadership remained unfazed and unperturbed.

On 27 April 2011, Nielsen claimed that the R&AW and the British Security Service (MI5) had jointly planned the arms drop operation so that violence could be precipitated in West Bengal, which could then be used as a pretext by the Congress regime to topple the Marxist government of Bengal Chief Minister Jyoti Basu. In April 2011, the central government's glib statement rebutting this, broadcast live on one Indian television news channel, was perfunctory, to say the least. Laced as it was with bureaucratese, it bespoke a culture of official lassitude and indolence, if not outright incompetence.

'The case remains under investigation. Any new facts emerging at any time will be looked into by the CBI in a professional manner... Government and the CBI are acutely conscious of the fact that nothing should be done at this stage that will prejudice the extradition proceedings or the intended trial of Kim Davy after his extradition to India,' part of the statement read.[6] Here was an excellent opportunity to reopen the entire case, bring the absconding Ananda Marga monks to justice and aggressively pursue several loose ends in the United Kingdom (UK), the United States (US), Hong Kong and Singapore where, one former CBI officer told me in October 2012, a prime accused, Deepak Manikan alias Daya M. Anand, the man who had brought the three parachutes from Johannesburg to Karachi and was subsequently the subject of an Interpol Red Corner Notice, continues to live freely. A small task force could have been set up, dedicated to concentrate on critical areas of the investigation and unmask the plotters and the end-users. Instead, nothing was done. Nothing at all is being done. Despite claims to the contrary, the government is letting sleeping dogs lie.

The official inertia and Nielsen's claim, outlandish though it sounded, began to sow some doubts in my mind. I asked myself disturbing questions time and time again: was there an R&AW hand in the arms drop? Were bit politicians such as Rajesh Ranjan, alias Pappu Yadav, of Bihar really involved, howsoever obliquely, in bringing in the weapons, using Nielsen as the courier? These questions led me to check and cross-check with different sources within and outside the establishment, whom I had never approached or spoken to before, in search of information for this book. Again and again, the sheer volume of information the sources revealed and the arguments they presented were contrary to Nielsen's claim. Some of these officers said, jocularly, that the R&AW would have done the security apparatus proud if indeed it had carried out the complex and perplexing Purulia arms drop operation with the exquisite finesse that was expected of an effective intelligence agency. If anything, the fragments of information that emerged from close scrutiny of files and a series of interviews with hard-boiled intelligence officers pointed to two possibilities: the Purulia arms drop was planned and executed by a gang of international smugglers or was a covert operation by a foreign intelligence agency. There was even a third possibility: the use of smugglers, sometimes referred to as 'dirty assets' or cut-outs[7], by a foreign intelligence agency to carry out its underhand tasks. Although the CBI and the R&AW did embark on the path to probe the hand of a foreign agency, it led them a certain distance, beyond which they encountered virtually non-negotiable mazes. Some officers asked the right questions: why were the Danish authorities so protective of Nielsen? What did they know that they did not want Indian intelligence or the CBI to find out? What precise role did Nielsen play in his dealings with the Sudan People's Liberation Army (SPLA) which, in the nineties, was being backed by the American Central Intelligence Agency (CIA) with lethal as well as non-lethal military supplies? Who really supplied the funds used to procure the weapons and the aircraft? How far did the Purulia conspiracy extend? But the case was so deep-frozen by the summer of 2011 that few officers had the zeal, the investigative acumen or even the stomach to pursue these gnawing questions, leave alone unravel the mystery. The CBI was not prepared to force the issues through. The agency shied away from traversing the full spectrum of the investigation, including Nielsen's suspected links with members of the Irish Republican Army (IRA), the drug cartels of South America and perhaps even the German terror group Baader Meinhoff.

Nielsen's televised misinformation had, fortuitously, given the Indian authorities another opportunity to take a final crack at the case. It went a-begging. The refusal by the Danish judiciary to extradite Nielsen to India, based on the unjustified and contrived fear that he would be subjected to torture and his human rights would be violated during incarceration in an Indian jail, should ideally have infused a fresh enthusiasm among senior CBI officers to take the case to its logical conclusion and answer the two most vital questions that have bedevilled the security agencies for nearly two decades: whodunnit and who were the weapons intended for? Instead, the CBI appeared defeated and spent.

On a related note, the brouhaha over Nielsen's extradition drew the attention, if any, away from the absconding Ananda Marga monks.

As one former CBI officer, who was very closely involved with the investigation, bristled, 'What bloody steps have been taken in the last few years to arrest these monks, against whom there is no Interpol Red Corner Notice, which implies that they are perhaps still in hiding in India?'[8] By the winter of 2012, the government (the Ministry of External Affairs) had become so utterly callous and indifferent towards the case that it permitted the Bangladeshi Army officer, Major General (retired) Subed Ali Bhuiyan to travel to India to take part as a 1971 Bangladesh liberation war mukti joddha (freedom fighter) in an Indian Army function commemorating the fall of Dhaka and the unconditional surrender of Pakistani troops forty-one years ago.[9] Maj. Gen. Bhuiyan had signed the end-user certificate (EUC) on the basis of which the weapons dropped over Purulia were procured from two Bulgarian arms manufacturing factories. Time had eroded institutional memory, if there was any.

There are way too many unanswered questions which, taken together, would be a lasting indictment of the CBI and the country's security establishment. I believe that had more resources—material and intelligence—been applied and a significantly different approach taken, the Purulia plot might have been uncovered and the identity of the men operating in the shadows could have been exposed.

As I began writing this book in the fall of 2012, my head hummed with the different possibilities, most of which I eliminated as implausible explanations about the conspirators and the likely end-users, until finally there came a point when a shape of the answer, howsoever indistinct and blurred, began to emerge. As I read and reread the documents and spoke with my sources, a pattern formed, pointing to one inescapable conclusion: the Purulia arms drop was a covert intelligence operation in which several 'cut-outs' and layers of shell companies had been used to give it all the elements of deniability. Yet, it could not be given the label of 'the best fit' because there was no smoking gun evidence to point a finger at any particular agency which could be linked directly to the arms drop. It seemed that I was in a looking-glass world in which the shadows only lengthened and the real players lurked in the darkness. After all, both the international criminal underworld and the world of intelligence are 'professions' that rely on deceit and intrigue, and the actions of their adherents are almost always devoid of hard evidence. My research indicated that the Purulia arms drop was planned at least three years before its actual execution. Nielsen's two fake passports in the name of Kim Peter/ Palgrave Davy (he had adopted the identity of a four-week-old baby from Rotorua in New Zealand)[10], which were procured in Wellington in 1991 and 1992, seem to indicate that the plan first originated at least three years before the actual event occurred.

This book relies on two basic sources of information. First, and primarily, Indian and foreign government files, a bulk of them classified, and other official documents, including witness statements, available in the public domain when the Purulia arms drop case went up for trial in 1997. Indeed, the very foundation of this work is based on documents that were studied, analysed, synthesized and verified over a period of time. A significant minority of documents and files I accessed contained material on intelligence sources, emails exchanged between investigating officers and methods they employed, which are not central to the narrative, but I read them to ensure that some of the conclusions drawn in the book do not contradict the contents of the official and classified files.

The second vital source of information constituted a series of in-depth interviews with some of the key officials, retired and serving, who were/are involved in investigating the case, besides open-source accounts in newspapers and parliamentary debates in India, the UK and Denmark. Initially, before embarking on writing the first draft, I gave a thought to adopting a style that is best suited to fictional thrillers, but quickly abandoned the idea and proceeded with a more journalistic, non-fictional approach. It gave me the opportunity to pack in facts as I read them in official files and documents, which I had painstakingly gathered over the last eighteen years, and insights and anecdotes that my sources narrated at their homes, in their office chambers and in restaurants and cafés in Calcutta, Delhi and elsewhere across the country. In many cases, as the reader will observe, I have avoided naming some of my sources, either on specific requests for anonymity or for the simple, important reason of their security.

Even as I have made a genuine attempt to present facts and stitch them together to narrate a story, I cannot claim that I have successfully covered every aspect of this incredibly mystifying case. For one, I was hamstrung by a paucity of funds that prevented me from travelling to distant places—the UK, Denmark, and the US—where some of the clues lay and could have been pursued. Second, I operated solo, without any government institutional support. This book is the result of an entirely private initiative and years of passionate following of the case. It would have been extremely satisfying to solve the Purulia puzzle in its entirety, but the secret world does not operate in plain sight and it is insuperably difficult to penetrate the shadows. I recognize that in the end, the shadows, though not as dark as before, remain, clouding the full truth from emerging.

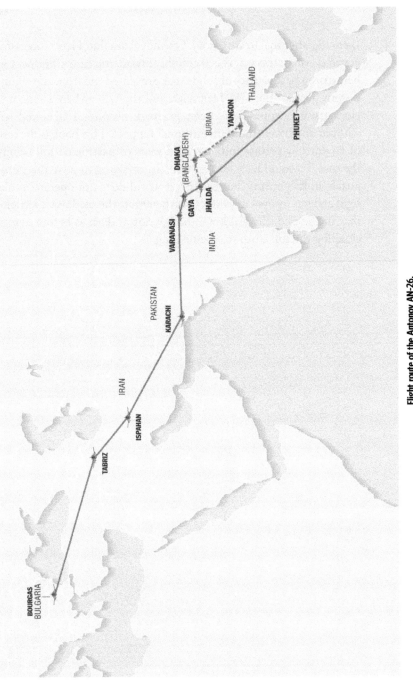

Flight route of the Antonov AN-26.

I

COLLECTIVE IRRESPONSIBILITY

25 November 1995. In his expansive wood-panelled air-conditioned room on the tenth floor of the dreary grey New Delhi headquarters of the R&AW at Lodhi Road, the joint secretary (security) read through the two-page note he had prepared after consultation with the agency's top management one last time. He scrawled his initials on it, called in his personal assistant and directed him to dispatch it by a special messenger. This is what he had signed on:

> It is reliably learnt that a Europe-based businessman had approached an individual with a request to pilot a small plane between Karachi and Dhaka, delivering arms and ammunition en route to 'communist rebels' in the area of Dhanbad. It was suggested to the pilot that he could land at an airstrip (86.20 degrees East and 24.30 degrees North) which is located along a river with hills on either side and close to the normal commercial air route from Karachi to Dhaka. After unloading the arms, the plane would go on to Dhaka, with its legitimate cargo.
>
> The individual concerned declined to take up the proposal. The arms dealer who was arranging the supply of the weapons is also learnt to have since pulled out of the deal on commercial grounds. Despite these set backs, the businessman who approached the pilot has since purchased an AN-26 aircraft. He left for Riga, Latvia, on November 15, 1995, and was planning to travel to Pakistan. Hence, there is a strong possibility that he is still pursuing the project.

According to further information which, however, requires corroboration, the insurgent group, which was to receive 2,500 AK-47 weapons and 1,500,000 rounds of ammunition of Chinese make, is opposed to the CPM government in West Bengal. The plan was to airlift the consignment of arms and ammunition to Karachi for storage by the side of the airport. From there the arms were to be placed on board a small aircraft which would have fled a route to Dhaka but would make a brief unscheduled landing at a rough airstrip in an area called Panchet Hill near Dhanbad.

Our enquiries reveal that there is a vast expanse of flat land near Panchet Hill which is located about 5 kms east of Netaria police station, Purulia district, West Bengal. This used to be utilised as a fair-weather airport some 20 years back by Calcutta companies having an interest in Dhanbad collieries. This location is about 9 kms south of Panchet village and 8 kms east of Chirkunda police station in Dhanbad district of Bihar. The place is sparsely populated but for about 400 Mahto tribals.

The information contained in this report may kindly be safeguarded in view of the sensitivity of the source.[1]

Nearly nineteen years ago, sometime in the middle of November, when the MI5's Security Liaison Office (SLO) shared this cryptic information with an R&AW officer,[2] it was considered an intelligence coup. The intelligence, though patchy and incomplete, was considered reliable and accurate enough for a senior R&AW officer in New Delhi to prepare a bowdlerized 'unofficial' (UO) note. On 25 November 1995, copies of the note were shot off to the then director of the IB, the cabinet secretary, the home secretary and the defence secretary.

Less than a month after the Indian security establishment learned of the conspiracy to smuggle arms into the country, it rained weapons from the skies over the nondescript and parched villages of Khatanga, Belamu, Maramu and Barudih in West Bengal's Purulia district, 250 kilometres west of Calcutta. The names of these villages did not figure in the R&AW's 25 November 1995, report shared with some of the senior-most bureaucrats and intelligence officials of the country. The weapons and ammunition were discovered by villagers early on the morning of 18 December 1995. At the time Indian security was in deep slumber, unaware that late on the night of 17 December, 'three containers containing more than four tons of sophisticated arms and ammunition were slung out under freight parachutes'[3] from a Russian-built transport plane, an Antonov-26. Where the R&AW had triumphed in gathering and quickly disseminating the intelligence it had received (from the Q Branch[4]) by way of working in liaison with the MI5[5], the IB and the Ministry of Home Affairs (MHA) had lamentably and miserably failed to act on the information supplied to them. The CBI later revealed that the IB and the MHA had done nothing at all to coordinate with other ministries, agencies and the administrations of West Bengal and Bihar to prevent a breach of the country's security. Enormous resources have been committed for decades to protecting the country's security but without seeking accountability of the organizations that are charged with doing so. The MHA and the IB displayed Neolithic incompetence, leading to quite a national security flap.

Three years into its yet-to-be-completed investigation, the CBI prepared a classified note to identify, establish and hold responsible government agencies and senior officers for their lapses that led, by all accounts, to one of the most bizarre and, admittedly, a spectacular operation to breach India's security. The CBI justified its indictment of the agencies involved—the Cabinet Secretariat, the IB, the MHA, the Directorate General of Civil Aviation (DGCA), the Ministry of Defence (MoD), and the Indian Air Force (IAF)—but at no point in time did successive governments of the Congress, the coalition of Third Front political parties or the National Democratic Alliance, led by the Bharatiya Janata Party (BJP), take steps to prosecute the individual senior officers involved. These officers, by their acts of omission, if not commission, were guilty of not acting in the interest of national security which, by the very positions they held, they were duty-bound to protect. Instead of being summarily cashiered and prosecuted, they continue to draw pension.

The story of individual and collective failures must be told first. It is in the context of those failures of an epic scale that the Purulia arms drop case—from the very beginnings of the conspiracy to the execution of the operation, the escape of the mastermind Niels Christian Nielsen from Mumbai's Santa Cruz (old) airport and his continuing residence in Denmark—can be located and narrated. It might seem that I am unjustifiably advocating assigning individual blame, but that is precisely the aim. More often than not, Indian officials have historically been exonerated or escaped punishment, even though there has been copious evidence to indicate widespread failures related to national security. The Purulia arms drop case exposed fault lines among senior and supposedly responsible officials within the government; the dysfunctions between the security agencies; the pervasive problems of managing and sharing information across large and unwieldy bureaucratic and intelligence structures which clearly had not shaped their policies, plans and practices to confront new threats and dangers arising out of the proliferation of small firearms in the wake of the disintegration of the Soviet Union; and the large-scale chaos in some of the Warsaw Pact countries after the demise of the USSR.

Lack of Coordination

The unsigned, undated CBI note given earlier shows that there was no unity of purpose between the concerned agencies. All of these agencies and the officers in-charge failed to meet the challenges posed before them between 17 and 21 December, when the AN-26 aircraft was made to force-land at Sahar International Airport while on its return to Karachi. Between themselves, the agencies pulled off a stunning failure. According to the CBI's report, the R&AW's UO note was addressed to the then Cabinet Secretary Surendra Singh 'by name' and yet no action was initiated by the Cabinet Secretariat, which oversees, at least, the administrative functions of the external intelligence organization. When CBI officers sought the Cabinet Secretariat's response, long after Surendra Singh had retired, a joint secretary, S.K. Mishra, claimed that the Cabinet Secretariat did not proceed with taking any action on the R&AW's UO note because it was addressed to the director of the IB. According to him, the IB was the appropriate agency to take suitable action against the threat.

Mishra went on to explain that since internal security was within the MHA's domain, it was the MHA's responsibility to act on the piece of intelligence. Clearly, the Cabinet Secretariat did not do its best, nor did its ranking officials set out any clear guidance for either the IB or the MHA.

The CBI's observation was that the explanation provided by Joint Secretary Mishra was not satisfactory, inasmuch as the UO note had given near-accurate details about the smuggling of weapons into the country from a foreign source with help of an aircraft. Besides, it was the CBI's contention that the intelligence was serious and credible enough for the Cabinet Secretariat to assume the role of a coordinating agency for various ministries to take remedial action. The CBI asserted that the arms drop took place despite the R&AW's information given to the cabinet secretary twenty-four hours earlier, showing a lack of response from the Cabinet Secretariat.

It was discovered, too, that the MHA, which indeed is responsible for setting guidelines and policy frameworks for the country's internal security, was not just impotent in the face of the excellent intelligence that was shared with it but had completely failed to perform the very basic task of alerting the two state governments—in this case, Bihar and West Bengal—whose border area was the suspected target or drop zone of the arms and ammunition. Among other top bureaucrats and officials, the UO note was addressed to the director, IB, with a copy to then Home Secretary K. Padmanabhaiah. The CBI's investigation disclosed that after receiving the R&AW intelligence on 26 November 1995, Padmanabhaiah made the following noting on the top-secret file, 'I hope DIB (director, IB) is sensitising Bihar state authorities.' Thereafter, the note was marked to a special secretary and three joint secretaries and further down to the lowest rung in the MHA's bureaucratic hierarchy, an undersecretary. One of the joint secretaries in charge of internal security, Shashi Prakash, on receipt of the note, sent out 'demi-official' (DO) letters by registered post to the chief secretaries to the West Bengal and Bihar governments on 12 December 1995, i.e. less than a week before the arms drop took place over Purulia. While responding to the CBI's queries later, the West Bengal government informed the federal investigating agency that it received the registered letter on 18 December 1995, but action could be taken only as late as 26 December 1995, by which time the arms drop had already occurred and one British and six Latvian nationals had been arrested in Mumbai. Sheer ineptitude on the part of the MHA, and the latter's incompetence to act quickly on the input, cost the country's national security dear. The Bihar government, however, informed the CBI that it never received the MHA's DO letter! Not only did home ministry bureaucrats display their foolhardiness, they did not settle for disaster when calamity could be found instead.

As a senior official of the lead ministry responsible for executive decisions on internal security, Joint Secretary Shashi Prakash explained his sending of the DO letters to the CBI. He also claimed, perfunctorily and casually, that the sensitive intelligence report was sent to the two state governments by registered post in accordance with the instructions laid down by the Manual of Departmental Security Instructions. For Prakash, it was a routine, bureaucratic chore and the information contained in the R&AW UO note was not something that troubled him to act beyond the confines of some archaic instruction manual. His action revealed that he had not even read the contents of the note, much less comprehended the implications of a possible ingress of India's airspace by a foreign aircraft laden with hundreds, if not thousands, of AK-47 rifles and ammunition.

There can be no two opinions that the intelligence shared by the R&AW was 'extremely sensitive' and actionable from the point of view of national security. As the executive department mandated with drawing up policy guidelines and plugging internal security vulnerabilities, the MHA should have promptly initiated steps to ensure that a breach of the country's security did not occur. The CBI observed, correctly, that the director, IB, was not independent of the MHA and, as the controlling ministry, it was the MHA's duty to monitor and regulate action on substantial intelligence inputs. It was also 'intriguing', the CBI pointed out, that action on a UO note was taken as late as 12 December 1995, sixteen days after it was received by the top mandarins in North Block, the mammoth colonial-era building that houses the MHA. This inaction on the part of the MHA was further compounded by the fact that the R&AW's intelligence was shared with the governments of West Bengal and Bihar by registered post, which naturally took some time to reach the authorities, and in case of Bihar, supposedly did not reach at all. Valuable time was lost. The MHA could have used more rapid means of communication—secure fax and/or telex lines, the IB's advanced, encrypted communication system, among others—to alert the state governments.

It was not just the CBI which made scathing remarks and observations on the ostrich-like attitude and functioning of India's security establishment. A bipartisan Parliamentary Committee on Government Assurances on the Purulia arms drop case 'found alarming shortcomings in the functioning of [these] governmental agencies in respect of sharing of important intelligence information, operations of unscheduled flights in [the] Indian skies, radar surveillance and, above all, inter-departmental coordination.'[6] More specifically, the committee held the MHA guilty for 'not conveying' the R&AW intelligence report to all the agencies concerned. 'Neither the Ministry of Civil Aviation nor [the] Directorate General of Civil Aviation (DGCA), which have been assigned the task of clearing all the flights, including unscheduled civilian flights, were informed of the intelligence report.'[7] The committee not only questioned the weak and non-existent command and control system and the lack of inter-ministry and inter-agency communication, but pointed to a 'major lapse on the part of the Ministry of Home Affairs'. So,

> [...] when the Air Traffic Control at Mumbai ordered the landing of the aircraft at Mumbai airport, no security man was found anywhere around the time of landing of the aircraft. A jeep came in which the driver was the lone man present and Mr Kim P Davy [alias of the mastermind of the Purulia arms drop operation] got into it and then disappeared. He is the main accused and he has not been traced so far.[8]

It appears that the committee was supplied only partial information (as we shall see later) on the sequence of events that occurred. After the AN-26 landed at Mumbai's Sahar International Airport, Davy was suspected to have boarded a jeep alongside two Airports Authority of India (AAI) officials who innocently escorted him to the terminal, from where he made a few phone calls abroad and within the country before making himself scarce.

Even with the spotty information provided by the various ministries and agencies to the committee, the parliamentary panel tore into the defence of the representatives of the organizations who made their submissions before it. 'The representatives of the ministries of Home Affairs, Defence and Civil Aviation who appeared before the committee to give evidence showed uncanny skill in blaming each other for not being able to prevent the arms dropping by a foreign aircraft. Each ministry tried to justify its inaction and said that it was the job of the other ministry,' the parliamentary panel observed.[9] The tapestry of lies, lapses and shoddy security work lay exposed.

After carefully considering the facts placed before it, the committee was, however, appreciative of the 'frank admission' by Home Secretary Padmanabhaiah that 'there was total failure on the part of the governmental agencies and that there was total lack of coordination among them.'[10] According to the committee's report, Padmanabhaiah admitted that information available to the government suggested an aircraft would drop arms around the date on which the actual arms drop took place. The area specified was Dhanbad in Bihar. Padmanabhaiah took the plea that although the likely place indicated by the IB was Dhanbad, the actual arms drop took place in Purulia in West Bengal. In his defence, the home secretary claimed that 'in a case like this, intelligence information cannot always be precise[11].' The committee conceded that 'certainly, intelligence information can point to only probabilities in such cases. But it was the duty of the Home Ministry to alert all the agencies' and direct them to 'look for a plane loaded with arms so as to intercept it before it fulfilled its mission. Instead, the Home Ministry alerted the Defence Ministry, the government of Bihar, and later sent a routine letter to the government of West Bengal informing them about the intelligence report[12].'

Cover-up

The mutual blame game and cover-up by the internal security machinery—of its failure to detect and track the AN-26 aircraft and prevent the arms drop—began soon after the discovery of the weapons consignment. In response to questions raised by a government-appointed committee of bureaucrats headed by then MHA Special Secretary V.K. Jain,[13] which recommended measures to coordinate the efforts of all agencies and departments to prevent any future violation of Indian airspace, the R&AW lashed back at the bureaucratic perfidy. A hard-hitting note of the agency found 'objectionable' the suggestion that the Ministry of Civil Aviation and the Department of Revenue should have been treated as the 'principal consumers of the information related to the Purulia arms drop[14].' The R&AW's interpretation—correctly—was that treating the Ministry of Civil Aviation as 'principal consumers' of the intelligence it had obtained from the British Security Service 'implies that the Purulia incident was a mere violation of the Indian Aircraft Act and the Customs rules, whereas it was very clearly an act of war directly affecting the internal security of the nation, particularly that of the states of Bihar/West Bengal where, according to our report, the landing was to take place. The information related basically to a matter of internal security about which the Home, Defence and IB were principally concerned. We did not view the information as being one of violation of revenue-related regulations[15].'

The R&AW's argument was that the information it had received was from

> [...]very sensitive quarters and a wide dissemination could have led to premature leakage as has been evident in the subsequent publicity and attributing the information to R&AW... The information required to be evaluated in the light of the ground situation and a coordinated action was [sic] plan was required to be formulated by those concerned with the maintenance of internal security. It was at that stage the Civil Aviation, Revenue, state governments etc were required to be alerted on a need to know basis.[16]

Before deciding to share the intelligence with the three principal consumers, the R&AW management had wrestled for a few days with the question of whether to share the MI5's information before it was actually passed on to the Cabinet Secretariat, the MHA and the IB. The then R&AW special secretary, Ranjan Roy, the second of the two deputy chiefs, finally opted in favour of sharing the intelligence. Although some senior R&AW officers found it hard to believe that an aircraft could possibly sneak into Indian airspace to drop weapons with the aid of parachutes, Roy, who was to later become R&AW chief, went with his gut feeling that even though MI5's intelligence was carefully worded, it was actionable.

Short of describing the committee of bureaucrats' stand as a deliberate effort to cover up the failure of the home ministry, the R&AW said, 'The report [draft report of the committee] has not gone into the question of why the government of West Bengal was not informed, which was one of the first preventive measures that should have been taken since we had identified Purulia as one of the likely landing sites.'[17] Unsparing of the committee, the R&AW note found it strange that the panel's draft report had failed to highlight the fact that the Ministry of Defence 'did not alert Air Defence Centres about the suspect AN-26 aircraft.' One redeeming feature of the Jain Committee was that it at least admitted—insipidly—that 'the incident relating to the para-dropping of arms and ammunition over Purulia manifested a weakness in the existing system and, in the interest of national security, there was a clear need for working out mechanisms which would ensure that such incidents are not repeated in the future.'[18]

MoD and IAF Culpability

If the MHA failed as a coordinating ministry, the MoD and the IAF brass were no less culpable. In the course of its investigation, the CBI stressed on the simple but critical evidence that the R&AW UO note was marked to the then defence secretary by name. As the MoD did not contact the civil aviation authorities, the DGCA permitted an AN-26 transport aircraft to fly over Indian airspace. Had the MoD sensitized the DGCA about the aircraft, as the R&AW intelligence report mentioned, the civil aviation regulatory authority would have been circumspect in granting permit to the plane. The CBI claims that no ministry other than the MoD had the wherewithal (in terms of the IAF's air power) to deal with the threat of a mysterious aircraft flying into Indian airspace to para-drop lethal weapons. The CBI found that the MoD had not issued any directive to the IAF and other authorities concerned in this regard. Had it done so, the airdrop could, perhaps, have been averted.

As per the Indian bureaucratic tradition of buck-passing, the defence secretary in his deposition before the Parliamentary Committee on Government Assurance took the stand that since the plane landed at the civil airport, the IAF had nothing to do with the security arrangements. Further, according to the committee, 'As per the evidence, the MoD had prior knowledge' that an aeroplane might intrude Indian airspace and try to drop arms in the border region of West Bengal and Bihar.[19] But the MoD neither acted on nor shared the information with the DGCA, leading the committee to observe that it was 'not convinced by the attempts made by the defence secretary to exculpate the IAF in this matter.' While it is true that the IAF did not have radar surveillance throughout the country at that time and it was not its function to monitor civilian aircraft criss-crossing the country's airspace, it was in a definite position to coordinate its efforts with the DGCA to track down the unscheduled flight. 'The fact is that it did not care to act on the intelligence provided to it... The MoD [tried] to minimise the seriousness of the Purulia arms drop,' the committee noted.[20]

While recording the oral evidence of senior MoD officials on 19 March 1997, the committee asked then Defence Secretary T.K. Banerji (who had succeeded K.A. Nambiar, the defence secretary when the arms drop took place), whether it was proper for the ministry to simply grant air defence clearance (ADC) to the AN-26 aircraft solely on the basis of DGCA clearance and without conducting any inquiry, especially in light of the R&AW warning. Secretary Banerji replied that the regulation of authorized civil air traffic across India was primarily the function of the civil aviation ministry. All types of civil aircraft that fly over India report to civil aviation authorities who acknowledge that and grant them permission. Banerji claimed that on the day and at approximately the time the AN-26 took off from (presumably from Varanasi to Calcutta), a flight plan was filed (again) with the civil aviation ministry and a copy of it was given to the IAF's military liaison unit (MLU) for air defence clearance. The idea, according to Banerji, was to convey to the country's air defence system that a particular flight had sought permission which was given. Therefore, the presumption ipso facto was that it was a legitimate flight.

The CBI, however, found during investigations that the AN-26 aircraft took off from Varanasi and landed in Calcutta without ADC from the IAF's MLU. Even on its return from Phuket to Madras (Chennai), no ADC was given by the IAF for the aircraft and it successfully landed at Madras's Meenambakkam Airport without an ADC number. A few days after the arms and ammunition were discovered in Purulia and the Indian police and the CBI began investigation into the circumstances leading to the weapons drop, a team of IAF officers visited the four villages for an on-the-spot inquiry. After analysing all the flights which had entered Indian airspace, the IAF officers identified the suspected aircraft with the call sign YL-LDB. A general alert about the Russian-built cargo plane was issued on 21 December 1995. However, in spite of the alert, IAF-MLU Wing Commander R. Radhakrishnan, who was based in Madras at that time, failed to detect the aircraft that evening when it landed at Madras airport for refuelling while flying in from Phuket en route to Karachi.

None of the Members of Parliament (MPs) on the committee had any specialized technical knowledge, but they asked penetrating questions on the IAF's surveillance radars and why they could not detect the intruding AN-26 even when it crossed one of the most militarized border regions (Pakistan–India) in the world. The panel was most curious 'about the present position regarding (air) surveillance and radars in the country[21].' To pointed questions, T.K. Banerji explained that the IAF possessed two types of radars:

> There are radars which are used at air force bases for controlling the air space, regulating flight operations and for

exercises involving routine surveillance in the neighbourhood of the IAF stations. The second type includes air defence radars installed along threatened (international) boundaries. These are positioned in areas where, at the time of hostilities, they could be used for air defence systems or air defence alertness. These radars are not positioned country-wide, but installed in specific areas.[22]

When some of the committee members sought to know whether the surveillance radars operated round-the-clock, an IAF representative, Air Vice-Marshal (AVM) M. McMahon, claimed that 'it was not possible to keep the radars turned on for twenty-four hours for the reason that they would actually burn out.'[23] Banerji admitted that the IAF did not have any surveillance radars in southern India at the time,[24] as a result of which the AN-26 could not be detected before it landed in Madras on the night of 21 December 1995.

Director, IB's Failure

The R&AW's UO note was directly addressed to the then director, IB, Dinesh Chandra Pathak, but the country's vaunted internal intelligence agency, with substantial financial, communications and manpower resources at its disposal, failed to take any action at all, either before or immediately after the arms drop. First, the immigration authorities at all the domestic and international airports in the country were not alerted. Had the IB shared the information of an imminent danger with the Bureau of Immigration—which is but a branch of the internal security agency—it was likely that those who planned and executed the arms drop conspiracy could have been apprehended when the aircraft landed at Varanasi airport for refuelling. Besides, as the immigration authorities in Mumbai had no prior information about a suspect foreign aircraft and had no inkling that the aircraft had been force-landed, the principal accused Kim Davy could manage to escape from the airport terminal. Second, it is the CBI's case that the 'IB did not appear to have passed on the R&AW intelligence to the governments of West Bengal and Bihar. Had this basic measure been taken, the Calcutta airport security and immigration officials would have been on alert and the AN-26, which landed at Dum Dum airport in the early hours of December 18, 1995, after airdropping the huge consignment of weapons and ammunition over Purulia, could possibly have been intercepted.' Third, the CBI found during investigation that there was some activity on the part of the IAF, if only of a bureaucratic and diffuse nature. Then AVM V.G. Kumar, who was also director of air intelligence, informed director, IB, Pathak over a secure RAX (restricted area exchange) line of the decision to force-land the AN-26 aircraft over Mumbai, as it was suspected to be involved in the arms drop. In a written response to the CBI's questionnaire a couple of years later, AVM Kumar said the information that the IAF had decided to scramble two fighter jets to force-land the intruding aircraft was shared with Pathak at approximately 0040 hours on 22 December 1995. This fact was later admitted to by Pathak in his response to the CBI in his DO letter No. DCP/Pers/1998 of 8 July 1998, addressed to Loknath Behera, the then Calcutta-based superintendent of police (SP) of the CBI's Special Crime Branch. Having shared the information with Pathak, AVM Kumar claimed to have continuously kept in touch with the director, IB, and requested him to make adequate police arrangements at Mumbai airport so that the crew of the suspect aircraft were not able to escape. Around midnight, AVM Kumar contacted MoD Joint Secretary (Air) Vinod Rai who, on behalf of the Government of India, approved the IAF's decision.[25] Rai is understood to have disclosed to the CBI that AVM Kumar had kept him informed about the message, which he passed on to the DIB.

On his part, what Pathak did as the chief of the country's putative premier intelligence organization was too little, too late. When the CBI sought his statement on the chain of events that followed immediately before and after the aircraft was grounded, then Mumbai Police Commissioner R.D. Tyagi said that he was informed by Pathak about the suspect aircraft only around 0235-0240 hours, i.e. two hours after the director, IB, received the information that the plane was over the Mumbai skies and an hour after the aircraft had actually landed at 0139 hours at Mumbai airport. Ensconced in his plush and sprawling New Delhi bungalow, the director, IB, was either asleep at the time or was unable to take quick and effective decisions expected of arguably the government's topmost secret servant.

That Pathak feet-dragged or froze as the enormity of the threat and the collective goof-up dawned over him is attested to by the CBI's relentless investigation of bureaucratic and official lapses, willing or otherwise. The investigating agency's probe revealed that Pathak contacted a subordinate officer, Subsidiary Intelligence Bureau (Mumbai) Deputy Director C.M. Ravindran, over phone between 0130 hours and 0140 hours, i.e. an hour after he received the initial information and shortly after the plane had been made to land at Mumbai airport. Pathak, supposedly, explained to Ravindran the need to force-land the aircraft and directed him to alert the immigration staff so that the entire crew could be detained. In its remarks on Pathak's abject failure, the CBI said that the director, IB, lost valuable time in alerting the Mumbai police commissioner and Ravindran. Consequently, the failure to order and put in place proper security arrangements facilitated Kim Davy's escape from the airport. In a sloppy effort to protect his hide, Pathak later explained to the CBI in a written response that it was incumbent on the IAF to take steps to isolate and cordon off the aircraft till a decision was taken about what course of action was to be followed. The CBI, however, stuck to its statement that the director, IB, lost valuable time in passing on information, due to which a foolproof and proper police bandobast could not be made in time.

'Davy's escape from Mumbai airport was the second blow for us. When the AN-26 was force-landed in Mumbai, the then Special Secretary, Ranjan Roy, called to congratulate me well past-midnight on 22 December 1995. My elation soon turned to extreme disappointment when Davy escaped after being virtually escorted out of the airport by some Airports Authority of India staff,' recalled the joint secretary who had first dispatched the 'UO Note' to the cabinet secretary, home secretary and the defence secretary. Still nursing the frustration and irritation over the manner in which his agency's intelligence was treated casually nineteen years ago, the joint secretary said that in the weeks before the information was shared with the principal consumers, he and his staff had verified with the IAF whether a foreign aircraft could invade Indian airspace surreptitiously. 'I was given to understand that there was absolutely no way for such an eventuality and yet the very thing happened,' he said.

Although the Parliamentary Committee on Government Assurances never got the opportunity to question Pathak or other IB officers, who took shelter under the convenient trope of secrecy, it raised valid doubts and concerns on the complicity, if not active connivance, of some of the top bureaucrats and officers. In its seventh sitting, on 19 December 1996, the committee, while examining three senior CBI officers, including its then director, Joginder Singh, pointedly asked whether any officer belonging to either the MHA, IB, DGCA, or the IAF were 'involved in the airdropping conspiracy[26].' I am not sure whether the attending committee members had on their minds the notion that some of the senior officers, against whom the CBI found instances of lapses, were criminally involved in the conspiracy with the actual masterminds with the aim of facilitating the arms drop. To the committee's question, the CBI's response was that it had found nothing conclusive against Indian officers about their complicity in the case, but that it would move to prosecute any official if it found even the slightest hint of connivance with the principal accused persons. The course of the CBI's subsequent line of investigation strongly indicated that it had either lost interest in pursuing the complicity angle against some of the senior officers or it simply did not find enough evidence to prosecute them or did not have the nerves to go after them. The CBI promised the committee that even if they found that no official was involved, it would still assess other evidence and instances of lapses and report to the chief vigilance officers of the respective departments, recommending departmental action. This never happened.

Implicit in the committee's concerns on official connivance were two unstated and, therefore, complex and sensitive questions: did a mole in the Indian security establishment work in ways to facilitate and assist the Purulia arms drop masterminds and thereby compromise national security? Were only non-state actors involved in the conspiracy or did former agents of a foreign state agency play a role that had the whiff of a carefully and meticulously planned intelligence operation? The CBI, as well as the Interpol, which assisted and coordinated a great deal of the investigation with the Indian investigating agency, have shown extreme—and understandable— reluctance to pursue these questions, even though they did arise in the minds of some of the investigators and some seasoned R&AW hands who have since retired from service or passed away. Some of my R&AW sources, whom I cannot name here, disclosed to me in the initial weeks and months of the CBI investigation that they suspected at least one senior MHA officer of acting (or not acting) in ways prejudicial to the interest of national security. Steps to probe that civil servant for any association he might have had with foreign intelligence agencies or the Purulia arms drop masterminds was imperative, especially because earlier in 1995, an IB joint director on secondment to the Ministry of External Affairs (MEA) was purged from service for his alleged closeness with the New Delhi station chief (a woman) of the CIA. The central issue was a lack of will. The CBI's top management was terrified that intensive inquiries against senior officers and bureaucrats might trigger disaffection in the higher echelons. It entailed interviewing powerful civil servants and, needless to say, there were risks of leakage and avoidable publicity that comes with all high-profile investigations linked to the country's national security.

All that the CBI scrupulously followed and brought before the trial court in Calcutta was material evidence against those actually accused in the arms drop case. In the weeks and months of 1996, 1997 and 1998 I would often beseech my CBI sources with uncomfortable questions on the likely role that a particular foreign intelligence agency might have played in l'affaire Purulia. Each time they would merely plead ignorance or flash a deceptive smile and say, 'We are trying to remain very focused on the evidence that has emerged and continues to emerge in several countries about the involvement of those who have been indicted, those who have been brought to justice and those who continue to remain absconders.' The more the CBI officers tried to sidetrack the questions, backed up by specific instances of a foreign intelligence role, the more I was convinced that the CBI wanted to steer clear of any controversy. Besides, expanding the scope of the investigation to bring under its scanner another state agency or its agents would raise untold legal and diplomatic hurdles which the CBI could very well avoid if it made pursuing the suspects any easier. I am certain that the CBI found itself in a dilemma: on the one hand it believed in the need to make intensive and extensive inquiries about the possible role played by a foreign intelligence outfit, and on the other it continued to fear being overwhelmed by this line of investigation, in addition to the regular criminal case that it was pursuing against the fourteen accused persons. 'The goal is to be realistic and fully aware of our own capabilities in bringing to justice and seeking conviction of those found guilty of waging war against India. There is no point in burdening ourselves with avoidable and unmanageable externalities,' a senior CBI officer who was deeply involved in the investigations told me in 1997.

What the CBI, the R&AW and the Indian and foreign press were unanimous about was the collective paralysis in the Indian security establishment after the R&AW shared the vital information with the concerned authorities. The arms drop operation demonstrated India's airspace 'vulnerability' and the 'transgressors, with unprecedented money power, access to [the] latest technology, organizational strength, manoeuvrability and scope for strategic alliances with other like-minded groups'[27] selected their theatre of action. Once the R&AW's intelligence, shorn of all identifiers of individuals and names of places abroad but containing enough substance to be described as near-accurate, was disseminated among the principle consumers, no ministry, department or agency was put firmly in charge of managing the case, leave alone working effectively to translate the intelligence input into action; in this case, identifying and preventing the aircraft from entering Indian airspace or grounding the plane before it could offload its deadly cargo. And once the aeroplane managed to intrude into Indian airspace with consummate ease and was able to accomplish its mission, neither of the principles assigned any responsibilities across the ministries, departments and agencies. In other words, there was neither joint action nor even a semblance of some level of integrated approach towards neutralizing a threat from the skies. The players were in position, doing or not doing their jobs. But who was directing the play that assigned roles to help them perform as a team?[28] Unfortunately, no one. The only men who acted—audaciously, surely, methodically, effectively and with catastrophic consequences— were those on board the AN-26 aircraft, and their mysterious handlers elsewhere across the globe.

2

IS IT A CIA SCAM?

Peter Bleach at the age of sixty-two, at Scarborough, North Yorkshire.

On the mildly chilly but bright morning of 17 August 1995, a Scandinavian Airline System (SAS) flight from Manchester landed at Copenhagen Airport at 0830 hours. Seated in business class, Peter Bleach put on his jacket, adjusted his tie and waited for the door to open as he thanked the blonde stewardess for the on-board hospitality. At Arrivals, William Roeschke shook hands with him before ushering him into a metallic maroon BMW diesel estate car. Of Danish descent but a resident of Munich, Roeschke began a casual conversation with Bleach, telling him that since Roeschke was holidaying with his wife in Denmark, he was requested by a Dane, Peter Jorgen Haestrup, to receive Bleach at Copenhagen airport. Besides, Roeschke, a business associate of Bleach, would be able to easily recognize him.

Roeschke pulled up by a small restaurant where Bleach and he had coffee. On their way to Haestrup's house, they did not discuss business at all. This was Bleach's first visit to Copenhagen and he was more interested in absorbing the Danish capital. The weather was excellent and that lifted Bleach's spirits.

Roeschke and Bleach reached Haestrup's residence around 1000 hours local time. It was an expensive detached property facing the sea, in Dyrehaven, Klampenborg, a very upmarket suburb of Copenhagen. The house was a few yards from Haestrup's office and, therefore, the address was the same as given to Bleach during an earlier telephone conversation. It was not situated on a large plot but had a boat mooring directly in the front, and was surrounded on three sides by a deck, which projected into the sea. Bleach saw a couple of new Jaguar and Mercedes sports cars, which were parked outside. He later learnt that the Jaguar belonged to Haestrup, who owned a company called A.E. Lundgren & Co and had business interests and possibly a residence in Hong Kong. Haestrup had previously told Bleach that he partially owned some Danish nightclubs and was in the international tobacco business, trading mainly with China.

Roeschke took Bleach into the house where he was introduced to Haestrup, who stood at 6'3" and had a pale freckled complexion, blue eyes, a round face, a small nose and thin lips. He was of heavy build, and had short blond hair. He appeared to be in his thirties. Haestrup, who was warm, though not friendly, showed Bleach around the house which, he said, he shared with his girlfriend Marlena and a large Irish wolfhound. They had no children. He took Bleach to the dining hall where there was another man who was introduced to Bleach as Brian Thune, a friend and business partner who owned the Mercedes parked outside. Thune was about 5'10", of slim build, with plain, short, mousy hair. He had a small moustache and a thin mouth, the corners of which turned down in a distinctive fashion. Although Thune was of Danish origin, both Haestrup and he spoke good English. Thune was found to be the owner of the Copenhagen-based Danacot Steel Corporation.[1] Haestrup brought in coffee and a Danish liqueur, which the four sipped before Roeschke left. It was time to discuss business.

> I produced copies of my contract and I was told that they liked the offer that I had made and were happy with the goods and their prices. They now wished to discuss the terms of the actual sales contract. Basically, my terms were: payment 100 per cent in advance by irrevocable, confirmed,

transferable Letter of Credit to be drawn on a prime bank
and capable of sight payment at counters of my nominated
bank, upon representation of shipping documents.[2]

Bleach wrote the above statement[3] on a computer in the solitary
confines of a CBI safe house somewhere in Calcutta, a few days after
he was arrested at Mumbai's Sahar International Airport, where the
AN-26 was ordered to land while flying in from Phuket, Thailand, on
the night of 21 December 1995. At the time it was on its way from
Madras en route to Karachi, four days after it had offloaded a massive
consignment of arms and ammunition ('goods') over five villages of
Purulia district.

It seemed that Thune was not happy with Bleach's terms and tried
to negotiate cash on delivery. Bleach remained firm and told Thune
that a cash transaction was impossible because the weapons factory
required payment immediately upon shipping the goods from its
premises. An unpleasantness began to creep into the nascent
relationship. 'This carried on for quite some time,' Bleach wrote
later.[4]

Thune suggested that Bleach should negotiate credit with the
factory which, the Briton exclaimed, was unheard of and ridiculous.
'Thune and I did not get on well and I assumed that he was one of
those businessmen who were simply difficult for sake of being so,'
Bleach said in his statement. Bleach assumed that Thune wished to
profit from a month or so of extra interest on funds he would receive
from his client. After an hour of intense and clamorous negotiation,
the matter was resolved, in that Bleach did not budge from his position
and both Thune and Haestrup had more or less agreed to the British
arms dealer's terms. During the course of the business meeting, Bleach
was aware of the presence of another person out on the deck outside
the house. He assumed it was Haestrup's girlfriend. That person—
not a woman—would slowly make his presence felt in subsequent
meetings, and much later during the execution of the clandestine
operation.

Thune asked Bleach how he would deliver the goods to Calcutta
port. Bleach said that he was yet to decide whether to use air freight or

sea freight, but his quote would cover both, and if there was any fund left over after freight was paid, it would be credited back to his company (Bleach had assumed that Thune was with A.E. Lundgren & Co). Bleach said that it would be necessary for Thune to provide him details of his shipping agent in Calcutta and that he would then consign the goods to his agent's bonded warehouse at Calcutta port. He pointed out that the warehouse would have to be certified, cleared and approved for explosives storage well before the goods were delivered from the factory to the destination. Familiar as he was with trading in small arms, Bleach told himself that delivery would be unusually fast. Rifles and rifle ammunition were stock items on permanent and continuous manufacture and there was no production lead time. A deal such as this could be turned around in under a month and would involve very little work.

Thune then sprung a surprise. He said that he wanted to know how Bleach proposed to deliver the goods *outside* Calcutta port. The wily Bleach immediately suspected that Thune was hinting at an illegal delivery of arms and ammunition. So he directly asked Thune if that was the case. Thune admitted as much. Bleach, to satisfy himself completely on the nature of the purchase and delivery, again asked Thune if he was being told to illegally deliver arms into sovereign Indian territory. Nodding his head, Thune replied, 'Yes, that is what it is, Mr Bleach.'[5]

'But this amounts to an act of international terrorism,' Bleach pointed out. Thune apparently agreed. Haestrup, sitting quietly at the table through the entire conversation, did not interject but was clearly a party to the discussion.

Caught in a dilemma, the Briton's mind began to race. Prior to entering the shadowy world of arms wheeling-and-dealing, Bleach had had substantial experience in the field of security investigations and counter-terrorism.[6] He thought he would have to report this entire episode immediately to the British authorities and, while he would be naturally disappointed in losing what had promised to be some good business, he had to find out as much as he could about the prospective weapons' buyers. He, therefore, decided to play along and go on with the meeting as though he was prepared to fulfil the order.

At the back of his mind ran the idea that he must report the facts to the British authorities once he returned to the UK.

Sensing an upper hand during the meeting, Bleach told Haestrup that he had been misled and that the prices and terms that he had quoted were strictly for a legitimate transaction. The arms dealer pointed out that the price quoted for freight (approximately $35,000) was also strictly for a legal consignment and, under the new circumstances, wildly inaccurate. With a pleasant smile, Bleach said that he would be quite happy to consider the buyers' proposal, but the prices would have to be revised upwards rather sharply. Both Haestrup and Thune said they would be agreeable to this and asked what sort of price increase was involved. Bleach said he would keep the unit (per gun and per round of ammunition) cost as it was but now he would quote separately for the acquisition of necessary documentation and specialized freight delivery. Documentation and handling fees would be at least an additional $50,000, payable in cash in advance, but Bleach added that he would not be able to give any kind of estimate for the delivery until he had more details of where the actual delivery point would be. Of course, he did not expect the buyers to disclose the precise delivery point at that stage of the discussions.

Piercing Blue Eyes

Adopting a pleasant demeanour, Haestrup and Thune suddenly led Bleach outside the house to the deck, where the British arms dealer was introduced to a casually dressed, athletic-looking man with blond hair and a slight beard on his sunken cheeks. He was wearing steel-rimmed spectacles behind which were a pair of piercing blue eyes. 'Although I was not told his name that day, I now know him as Kim Peter Davy [he did not then know Davy's real identity and name, which was later found to be Niels Christian Nielsen],' Bleach wrote in his statement to the CBI.[7] What was revealed, ominously, was that the man was in direct contact with the insurgent group on the ground in India and that he would provide more precise details about the point of delivery. 'Davy said that the arms were to be delivered to a group which was operating several hundred kilometres to the west of Calcutta,' Bleach claimed.

Kim Davy aka Niels Christian Nielsen (*left*) when he was barely out of his teens; and today.

What followed next at the meeting was a protracted but lively discussion on how best to deliver the weapons. A covert landing by boat on a remote stretch of the coastline was first discussed. However, this idea was abandoned because of the risks of onward transport across a large tract of Indian territory. Several trucks and jeeps would be required and it was considered that the chances of random discovery were simply too great. So Bleach suggested that the only really viable method of delivery would be by air, either by covert landing or by parachute drop. Davy, or Nielsen, quickly pounced on this brilliant idea. He said he would consult with the people on the ground and find out which would be the best alternative.

Quite obviously the man in charge, Nielsen told Bleach to calculate the various costs involved and liaise with Haestrup on the matter. During this conversation, an inquisitive Bleach asked Nielsen about the group concerned and tried to obtain more information. But Nielsen was no amateur. He only said that it was a non-political group, which was being oppressed by officials of the communist government of West Bengal, and the group members merely wished to defend themselves against this oppression. No names were given. For Bleach, this was not a particularly convincing explanation, especially because 2,500 AK-47 assault rifles was a very large consignment for a 'small group'. Bleach was more than sceptical;

nevertheless, he did not expect to be given much accurate information at this stage.

Of course, Bleach did not know that a few months before the Copenhagen meeting, sometime in March, Nielsen had visited the restive Peshawar region in Pakistan's North West Frontier Province and inquired about the possibility of buying weapons from the sprawling illegal arms bazaar there. For two weeks he scoured Darra Aadam Khel in Peshawar in search of suitable weapons. It is unlikely that he travelled to Peshawar all by himself for, by early 1995, the Pakistani town overlooking the Khyber Pass was a dangerous place for any European or American national looking for Kalashnikov rifles and other small arms. Peshawar, which had a US consulate, and other smaller towns on the lawless Pakistan-Afghanistan border, were in the grip of mujahideen terrorists backed by the Pakistani Inter-Services Intelligence (ISI). It was the time when towns and villages on the Pakistani side of the border were experiencing the aftershocks of the spectacular military victories of the ISI-sponsored Taliban sweeping across civil war-ravaged Afghanistan. Peshawar swarmed with Islamist Pashtun terrorists, Afghan refugees and ISI operatives. The plan to purchase weapons in Peshawar was abandoned because the price quoted for a second-hand Kalashnikov rifle was $150, which was rather high. That was when Davy decided to tap the Eastern European market.

After Bleach agreed to calculate fresh costs, Nielsen left. The other three—Haestrup, Thune and Bleach—then went for a late lunch at a nearby restaurant. As the lunch was about to end, Haestrup called Roeschke on his mobile phone and asked him to come over and drop Bleach at the airport. Over coffee at Copenhagen airport, Bleach expressed confidence that the deal would materialize and that he would keep in touch with Roeschke over the matter. He thanked Roeschke for the business recommendation but did not utter a word about the illegal aspect of the whole affair. He then disappeared through one of the gates to take the flight back to Manchester.

After reaching, the Briton took a train and reached home in North Yorkshire around 1230 hours GMT. As it was a Friday night, there was nothing much he could do until Monday morning. He heard nothing more from Haestrup in the next two days.

Meanwhile, a day after leaving Haestrup's Klampenborg house, Nielsen made a credit card (Visa) payment to Hertz Rent-a-Car in Copenhagen.[8] This was an innocuous payment of no apparent importance at that time, but would assume great significance when the Interpol and several other police and security organizations across a number of countries launched a manhunt to track his movements following his escape from Santa Cruz airport in the early hours of 22 December 1995.

The following day, on 20 August 1995, Nielsen wrote to 'Bryan' (his UK-based British lawyer, of whom not much is known), giving him instructions on how to proceed with the transaction, and seeking a contract plan. Davy said in his message that he had talked to people in India' and they wanted to know whether the person who would 'inspect the area (drop zone) could come soon so that they will not unnecessarily prepare a place that is not suitable'. In the letter, Nielsen mentioned a sum of $535,000 for '2,500 units' and added, 'I have lost $40,000 because of currency fluctuations on the DM [Deustch Mark] and Swiss Franc in SA [South Africa?], so my financier here is very anxious to see something materialize'.[9]

The Beginning

Sometime in May 1995, one of Brian Thune's contacts in Copenhagen, who went by an unusual nickname, 'Tommeren' (The Emptier), had told him that someone was interested in purchasing weapons. This was enough for Thune, who claimed to own Danacot Steel Ltd, which was registered in England and traded in commodities, to get interested in what he thought could be a lucrative deal. Tommeren gave Thune a Denmark cell phone number (+40-59-30-20), saying he could call up a Peter Johnson, the man who was interested in buying weapons. About a week later, Thune met Peter Johnson, who had crossed over from Sweden on a hydrofoil boat in Havnegade, Copenhagen. Although Johnson spoke Danish and said he had previously lived in Denmark, he was very reluctant to share more information about his Danish identity. Tommeren opened the meeting by introducing Johnson as living in Hong Kong, saying also that Johnson was interested in buying gold. Johnson claimed that he represented a large

Hong Kong-based conglomerate whose main interest was buying and selling gold upwards of 100 kilograms. Although the stated purpose of the meeting was to discuss possible ventures in gold, Johnson straightaway asked Thune whether he knew people who would sell weapons. At the end of the meeting, which lasted about an hour, Johnson and Thune exchanged phone numbers. Johnson subsequently shared with Thune a Copenhagen cell phone number and a landline number in Miami, where he 'sought refuge' whenever he felt he was under 'too intense scrutiny'.[10]

A few days after the Havnegade meeting, Johnson telephoned Thune, seeking another meeting at Café Zeleste in Nyhavn, Copenhagen. Since Johnson was already in Copenhagen, the two met the very day he called. Thune was a little surprised at the way the meeting turned out because Johnson was far more interested in weapons than discussing a possible business deal in gold. Almost in desperation, Johnson asked Thune whether he had made any inquiries in the small arms industry and kept insisting that a seller must be identified as quickly as possible. Indeed, Johnson cleverly let slip that he was interested in Kalashnikov rifles, which were to be delivered in India.

During his numerous meetings, which took place at various pubs, including Wilders at Christianshavn and Zeleste in Nyhavn, Thune 'established for a fact that Peter Johnson could raise considerable amounts of money, millions, by just lifting the receiver... Thune knew that Peter Johnson was also involved with drugs trafficking to a considerable degree... This was managed by a group in Hong Kong, but Thune was sure that some of it was earmarked for Denmark'[11].

Clueless about where to find a weapons' seller, Thune got in touch with his friend and business associate Peter Jorgen Haestrup who, he thought, would be able to help as he had contacts in Poland. At the time, Haestrup ran a firm called Inside-Outside (which dealt in anything from jugs and bicycles to aeroplanes), in Copenhagen's Birkerod area, in partnership with one Flemming Soderquist, who claimed to be a former chief executive officer (CEO) of the Hong Kong branch of the Danish pharmaceutical giant Novo Nordisk.[12] After a flurry of phone calls sometime in early June 1995, Thune and Johnson again met at Café Zeleste, where Haestrup was introduced to the 'buyer', Johnson, who now called himself Kim Peter Davy, from New Zealand. Haestrup had already made some inquiries with a Polish arms dealer, Arthur Liephardt, who had links with a retired army general representing a company called Fenix Metal. He now suggested that Davy and he visit Warsaw to discuss the modalities of procuring the Kalashnikov rifles and ammunition there. Four days later, the three were at Warsaw airport, where they were met by the general, who immediately took them to meet the dealers. Once the conditions were discussed, the dealer insisted on two crucial prerequisites—the money had to be given as soon as possible and Johnson had to justify why he wanted to purchase weapons, as well as furnish all necessary permits, including a valid EUC. Davy did not find the conditions unreasonable, exuding confidence that all of the paperwork would be done in no time. Although the place of delivery was not discussed, it was made clear to Davy that the weapons were refurbished. Bank information was exchanged and it was agreed that Davy would send across the documents within a few days. On the flight back to Copenhagen, a confident Davy told the others that the paperwork would be managed by his lawyers in Hong Kong. He also flashed his Hang Seng Bank gold card as to show that money would be no problem.

Less than a week after the Warsaw meeting when Thune, eyeing his share of the brokerage, called Davy, he was told apologetically that the project had to be dropped because the price quoted by the Poles was too expensive. Besides, Davy said, the papers were not ready. He urged Thune to look for an alternative seller. Thune was, however, not disappointed because he still thought that he would be able to make some money, especially when Johnson dangled the gold carrot. The two decided to fly to South Africa. Two weeks later, Thune and Davy were on a flight to Johannesburg. They travelled to Durban, where they met the Danish consul Per Bjorvig and explored the possibility of trading in gold. Despite a visit to Pretoria where they met a contact whose name was provided by the Danish council, the deal to buy 200 kilograms of gold did not materialize. The two left South Africa, Thune for Denmark and Davy for the US, where he most probably visited his ranch in Trinidad, Las Animas county, Colorado.

Brian Thune.

In meetings over the next one month, Kim Davy would talk to Haestrup about his work among poor Indian children and orphans who needed medicine and food, along with the cover story that the weapons would be 'for the government of an Indian state in [the] same area where he worked with children'[13]. At every subsequent meeting with Haestrup, Davy spoke in English. In one of these conversations, Davy shared his thought that since India lacked the infrastructure, he was interested in buying an aeroplane that would help transport medicine and food for the orphanages.[14] Haestrup, suspecting nothing at the time, contacted an old associate, William Roeschke, a Dane who lived in Munich, who suggested Peter Bleach of North Yorkshire in England as an efficient weapons' supplier.

The Deal Is Struck

In early August 1995, the landline phone at Bleach's Howdale Farm residence in North Yorkshire rang several times before he picked it up to recognize the voice of William Roeschke.[15] After the customary pleasantries were exchanged, Roeschke made the pitch that one of his business acquaintances had mentioned that he had secured a contract for the supply of some rifles and ammunition and that his colleague was not happy with the prices offered by various suppliers. Roeschke told Bleach, who was director of Aeroserve UK, an export company legally dealing in arms, that he had given his name and telephone number to his acquaintance as being a reliable and reasonable supplier and that he could expect to receive a telephone call from his contact soon. Of course, Roeschke said that should any business arise out of the introduction, he would expect a commission. Bleach laughed, thanked Roeschke and assured him that in the event of a successful business deal, he would indeed pay an 'intruder's fee' on a percentage basis. An upbeat Roeschke divulged his acquaintance's name as Peter Haestrup from Copenhagen, saying that he had worldwide business interests. Roeschke then hung up.

A few hours later, Bleach received another telephone call (at +44-1947-880-842). This time, the caller identified himself as Peter Haestrup and offered Roeschke's name as reference. Haestrup quickly came to the point: he had secured a control for the supply of 2,500 AK-47 rifles and 1,500,000 rounds of ammunition. He had already made some inquiries but he was not satisfied with the prices quoted. At this point Bleach jumped in, saying he could supply the same at very competitive prices and that his goods would be of Chinese origin, brand new and ex-factory. When Haestrup asked Bleach to provide him a written quote, the Briton said he would only do so after receiving an official written request on Haestrup's company's letterhead.[16] Haestrup also asked Bleach to include a commission for him in the overall price so that he could pass his quote unchanged to his client. Bleach agreed to this proposal, telling Haestrup that he would need to have some idea of the destination in order to quote freight accurately.

Shortly thereafter, Bleach received a fax from Haestrup on the Dane's company's official writing paper—A.E. Lundgren & Co—confirming his verbal request for a quote for 2,500 AK-47 rifles, together with 1,500,000 rounds of ammunition, 'C+F' (carriage and freight) Calcutta port.

Bleach prepared a quotation and faxed a copy through to Haestrup. He marked the file 'State Express'. The quotation sent to Haestrup

offered three options for the guns—AK-47 Type 56 Mk-1 at
approximately $80–$85 per piece; AK-47 Type 56 Mk-2 at $85–$90
apiece; and AK-47 Type 56 Mk-3 at $95–$100 per piece. He added
$2 per piece as commission for Haestrup. Ammunition (common to
all three variants) was quoted as standard military ball at $90 per
1,000 rounds (calibre 7.62 x 39 mm, as per standard AK-47). All the
goods were quoted as brand new, unused and ex-China factory with
standard six months' unconditional guarantee. Pre-shipment
inspection was available at buyer's expense.[17]

At this stage of the deal, Bleach felt that the goods that were to be
consigned at Calcutta port were not in any way significant. First, the
quantity involved was quite appropriate for a small to medium-sized
country, which might have been simply topping up its stock levels.
Besides, it would be a very reasonable quantity for a police or security
force. Second, he did not expect that the identity of the end-user
would be revealed to him, at least not until a firm contract for sale had
been properly signed. This would be normal business practice. The
same procedure would apply to the EUC, which he would not expect
to receive at this juncture and he would not ask for until a contract for
sale was signed. As far as Calcutta port was concerned, he presumed
that in this case, Calcutta port meant either a sea or an airport and was
simply a geographical definition, not an indication of the actual
location. There was no reason to feel suspicious about it because, in
order to disguise the identity of the end-user, at least in the early
stages of a contract, it would be quite normal and reasonable to
initially quote a port of destination which could be different from the
one intended, but located within the same geographical radius on the
same sea route. Later, though, Bleach would be mildly surprised on
learning that Calcutta port was indeed the destination.

There were two possible reasons for not disclosing the end-user
when the contract was still at its nascent stage. First, for pure business
reasons, the middleman (in this case, Haestrup) would not disclose
the end-user to the supplier (Bleach) as there was a good chance that
the supplier would simply approach the end-user directly and offer
the goods at a lesser price than the middleman. Companies or
individuals who trade together on a regular basis get around this

problem by signing 'non-circumvention agreements' (which is what Bleach and Davy did in their contract no. 10026/95/A of 13 September 1995) which effectively prevent all parties from approaching each other's business contacts directly. A second reason was that it was a contractual stipulation that the end-user's identity not be disclosed until the contract was finalized. This was particularly true when end-users purchased trial consignments of goods as it prevented the supplier from inflating the cost of the goods (when an initial order was to be followed up by a much larger one).

As far as pricing was concerned, the cost of basic items, such as the rifles, did not fluctuate very much. According to Bleach, he would never approach a factory for an accurate price at an early stage of an inquiry. The cost to the end-user would contain provision for commissions and profits and any fluctuation, up or down, would be absorbed within these provisions. For example, the cost of an AK-47 Type 56 Mk-2 (plastic butt, stock and pistol grip, complete with two magazines, cleaning kit and bayonet) from Poland would be around $135 per piece, and the same from China would be around $85 per piece, depending on the quantity ordered.[18] With Bleach's expertise in this field, he knew that prices had not changed for around three years and no special inquiries were, therefore, needed to provide an accurate quotation for Haestrup.

Several days later (on a Thursday), Bleach received a second telephone call from Haestrup at his North Yorkshire office to inform him that he was satisfied with the offer and would accept Option One (AK-47 Type 56 Mk-1 with wooden butt, stock and pistol grip). Haestrup asked Bleach whether he would be able to go to Copenhagen for a meeting to discuss the terms of the contract. Bleach said he would be happy to do so, provided Haestrup paid for his airfare and met all other expenses. For Bleach this was a tactical ploy, for he felt that if Haestrup was prepared to pay for his air passage, it would be an indication that his inquiry was genuine and, therefore, it would be worthwhile putting in greater effort.

Haestrup readily agreed to pay for Bleach's expenses and asked him where the nearest international airport was located. Bleach told him it would be convenient for him to fly out from Manchester. Haestrup promised to call him back.

Early the same evening, Haestrup did call, telling Bleach that he was in touch with the SAS office in Copenhagen and that he had purchased a return business-class ticket for Bleach for Manchester–Copenhagen–Manchester, valid till the next day (Friday). Bleach could collect the ticket from the SAS ticketing counter at Manchester Airport, from where the flight would depart at 0730 hours GMT the following morning. Arrangements would also be made for Bleach to be met at Copenhagen Airport.

Later that evening, Bleach prepared a standard contract for the purchase of 2,500 AK-47 Type 56 Mk-1 rifles, together with 1,500,000 rounds of standard military ball ammunition priced at approximately $390,000 C+F Calcutta port. Like a methodical businessman, Bleach printed several copies. The next morning he set off early for Manchester airport, carrying in his sleek briefcase copies of the contract, as well as a selection of leaflets and brochures in case there was an opportunity to discuss other business opportunities.

The opportunity to discuss other possible business ventures in Copenhagen never arose. The meeting with Davy (as Bleach knew Nielsen then), Haestrup and Thune was completely consumed by the order for the supply of a mammoth consignment of arms and ammunition that would fetch him a decent profit margin. Bleach was troubled by the fact that he was about to venture into an illegal deal whose implications he was yet to comprehend. He was not surprised that it was not legit, because illegal arms deals were an everyday affair. What surprised Bleach was that Davy and Co. had approached one dealer and asked him to carry out the entire delivery. Bleach had expected at least one 'cut-out' so that no one person outside the buyers' group would know the whole story. It would have been more sensible for the buyers to have asked Bleach to deliver to a bonded warehouse in a busy port such as Calcutta, where another agent would have taken over and delivered the goods to their final destination. Bleach began to wonder whether this might be a CIA scam, several of which had been carried out in Europe in the late eighties and early nineties. He decided that regardless of the deal being illegal or a scam, he would report to the British authorities that he had become privy to an illegal international arms shipment that was likely to be delivered to terrorists in India. He decided to play along with Nielsen & Co and organize the purchase of the arms and ammunition, while at the same time taking the first step to bring to the notice of British intelligence that an international cartel with deep pockets was about to finance, procure and clandestinely smuggle deadly weapons into a South Asian country.

3

LIKE VULTURES TO A CARCASS

The Purulia 'war package' was but a minuscule fraction of the Cold War legacy of massive stockpiles of SALWs. Indeed, the end of the Cold War, marked by the disintegration of the Union of Soviet Socialist Republics (USSR) and widespread political and economic instability in the Warsaw Pact countries of Eastern Europe, produced a 'volatile combination'[1] of surplus stocks in SALWs and lax or non-existent international regulatory mechanisms on their trafficking on a global scale. Millions of suddenly surplus weapons were sold at discounted prices as Eastern European countries tried to reduce stockpiles or modernized their armed forces. Globalization contributed to the dual problem by creating opportunities for the burgeoning illicit trade and the phenomenal rise in international organized crime, making the world more unsafe than it was during the Cold War, more conflict-prone and consequently more dangerous, especially for the people and governments of developing countries. The unchecked international trade in and transfer of SALWs, which was the most destabilizing feature after the Cold War, increased 'cultures of violence and impunity throughout the world'.[2]

There is no agreed definition of SALWs. However, the most widely used is the one proposed by the United Nations (UN) Panel of Governmental Experts on Small Arms, according to which SALWs include revolvers and self-loading pistols, rifles and carbines, sub-machine guns, assault rifles and light machine guns. Light weapons include heavy machine guns, handheld under-barrel and mounted grenade launchers, portable anti-aircraft guns, portable anti-tank guns, recoilless rifles (sometimes mounted), portable launchers of anti-aircraft missile systems (sometimes mounted) and mortars of calibre less than 100 mm. Ammunition and explosives include cartridges (rounds) for small arms, shells and missiles for light weapons, mobile containers with missiles or shells for single-action anti-aircraft and anti-tank systems, anti-personnel and anti-tank hand grenades, landmines and explosives.[3] These have, increasingly, played a greater role in conflicts globally, especially intra-state wars in which the weapons for the combatants came 'from outside, either because of ideological or other reasons'.[4]

For the 'new breed'[5] of arms dealers, traffickers and brokers, mostly operating out of Western Europe, the US and the erstwhile satellite states of the Soviet Union, and their 'natural partners'—international crime syndicates—the process of globalization, which had compressed space and time, and led to newer and advanced modes of transportation and communication, made the shipment of SALWs across continents relatively easy. By their very nature, SALWs, which in the early and mid-nineties were relatively inexpensive and durable, are easy to conceal in covert operations.

There were essentially four channels of small arms supply to state and non-state actors—government-to-government, commercial sales involving well-established global companies in the West, covert deliveries by governments and their intelligence agencies or shady arms dealers and black-market weapons deals. One distinct category of small arms suppliers, which prospered in the post-Cold War period, comprised the black-market dealers. With the demand for illicit arms growing in many war-torn areas, especially Bosnia and countries on the African continent, the world's black-market suppliers worked overtime to provide the desired weapons to combatants in several parts of the globe. As the Berlin Wall 'came down, so did the controls (on weapons transfers), and the market was thrown open to those who could seize the moment'.[6]

Arms Dealers and Brokers

The international trade in small arms was extremely difficult to monitor, let alone control, partially because of the role of middlemen—

brokers and shipping agents—who arranged for the sale and shipment of arms and associated military and paramilitary hardware, both new and old. Countries that were formerly under communist systems of government—Bulgaria, Poland, Ukraine and Romania, to name a few—no longer had the Soviet Union and its allies as the principal customers for their arms production facilities. 'The end of the Cold War made the bloated defence industry and large inventory of weapons in Russia and other former Warsaw Pact countries an easy mark for grey-market brokers.'[7] The countries, flush with copious volumes of SALWs, looked for new markets to supply their newly formed but economically hard-pressed governments with operating capital. This was accomplished by supplying arms and military material to any group—government or insurgent—willing to pay. The international arms dealers and brokers, who operated within a grey zone of legality, took advantage of these 'new' markets, supplying their lethal wares to criminal organizations and ethnic groups-in-conflict, for whom the weapons of choice were SALWs. Writing in 1999, Brian Wood and Johan Peleman, who were among the first Western European activists-scholars—perhaps the only ones—to use the Purulia arms drop case as one among several examples to highlight the arms dealers-brokers-shipping agents' nexus that operated in the early and mid-nineties, said:

> In today's complex global markets, more or less unregulated brokering agents can fairly easily arrange international arms transfers by bringing buyers and sellers together in an atmosphere of secrecy. The brokers do not necessarily buy and sell the arms themselves, but they make money from the transaction. The more controversial the deal, the more they go to considerable lengths to disguise the payments. Brokers have to make sure the arms are delivered, so they work closely in league with specialized transport agents. The latter make the arrangements by air, sea and road; if the cargo route is legally questionable or unsafe, the agent may engage in complex sub-contracting arrangements across several countries.[8]

By the early nineties, the global black-market trade in SALWs, 'almost always shrouded in secrecy'[9], expanded greatly and emerged as a major factor in the supply of arms to guerrillas, separatists, private militia and other non-state actors and sub-national groups. At the time, while it was impossible to put precise dollar values[10] to such trade, many black-market transactions ran into large quantities of arms and ammunition involving millions of dollars. In 1993, for example, British customs authorities intercepted a Polish ship carrying 300 Soviet-made AK-47 assault rifles, two tons of explosives and large numbers of pistols, grenades and bayonets destined for shipment to IRA extremists in Northern Ireland. Four years earlier, an Israeli arms company had smuggled 400 Galil assault rifles, 100 Uzi sub-machine guns and 250,000 rounds of ammunition to the Colombian drug lord Gonzalo Rodriguez Gacha via the Caribbean island of Antigua.[11]

The private weapons market served as an important—sometimes the only—source and chain of supply for militia, insurgents, criminal syndicates and 'informal clusters'[12] of terrorists. The illegal trade in weaponry flowed through transnational networks, as did narcotics and other contraband. Indeed, in many cases, the trafficking in narcotics was inevitably linked with the global trade and transfer of small arms controlled by different networks.[13] Many layers of individuals and places were involved, with some dealers being legitimate, abiding by the respective rules and regulations of their countries, while others were less scrupulous and operated in covert illegal arms sales.[14] Because the latter had to be secretly organized, the transactions that enabled them—deal-making, forgery, financial transfers, illicit transport—also had to be hidden and clandestine.[15] According to one noted analyst, the illicit trade involved 'three sets' of players: producers, recipients and traffickers. In this relationship, the two outer sets—producers and recipients—'rarely have any direct contact with one another'. The relationship is 'mediated' by the arms traffickers. This is the most prevalent procedure because the intended recipient is an insurgent group or a crime syndicate which is naturally 'barred' from procuring arms through legal means. In most cases, the recipient approaches the trafficker for 'assistance' in obtaining arms and ammunition. The trafficker then uses various forms of 'deception to obtain desired weapons from the (presumably) unknowing supplier. Once the arms are acquired, moreover, the trafficker arranges for delivery to the intended recipient, usually with the assistance of complicit shippers'.[16]

The expanding ranks of private arms dealers and brokers and their widely divergent backgrounds were a new development in the arms trade. One analyst argued that traditional arms merchants who were drawn from the ranks of largely retired soldiers, former arms company executives, ex-intelligence agents and government officials were joined by 'everyone from oil traders to toxic waste brokers to de-frocked Catholic priests' in weapons sales, thereby increasing the number of sources of weapons and facilitating their transfer and purchase.[17] SALWs were 'a major market for post-Cold War brokers, who played a key role in bringing together the arms stockpiles, parties to the conflict, and other interested parties'.[18] Once the Cold War ended, 'outlets' for small arms multiplied, commercial markets became 'differentiated' and the use of private intermediaries increased.[19] 'Stretching across a continuum of legality...to the shadow world'[20], the networks of arms dealers and brokers exploited legal loopholes in some countries, evaded customs and airport controls and falsified documents such as passports, EUCs, cargo papers and flight schedules. Illicit brokering activities in SALWs were, and continue to be, typically conducted through intricate arrangements involving complex transportation and opaque financial transfers.[21]

Intelligence Agencies

There was yet another source—as well as channel—of illegal weapons transfer during and after the Cold War: intelligence agencies, whether of Western, Russian, East European or South Asian countries. According to Wood and Peleman, these agencies used 'complex arrangements involving private airlines and clandestine shipments to disguise their operations'[22]—covert or camouflaged arms transfers to combatants in recipient countries. Needless to say, intrinsic to these private exports was the element of deniability, meaning that they were 'non-governmental and hidden'[23] from the rather extensive bureaucratic process designed to control arms trafficking. When non-state actors were supplied, as was often the case, it meant undercover operations, for which extensive networks were put in place to allow for deniability.[24] Arms dealers and those engaged in illicit brokering moved easily from one country to another and operated through intricate international arrangements, which made it difficult, if not impossible, to trace their activities and collect evidence to support the efforts of law enforcement institutions involved in investigating these deals. As Wood and Peleman observed, arms dealers and brokers and their gun-dealing activities had some common characteristics. They were:

- businessmen with military and security backgrounds and contacts
- motivated by economic gain rather than strategic political considerations
- able to use loopholes and enclaves of weak regulation between national legal systems to conduct 'legal' but often unethical business via third countries
- able to use agents and techniques developed in the modern international transport industry to conduct covert deliveries to sensitive destinations
- able to arrange complex international banking transactions and company formations in many countries, including the use of tax havens
- able to locate sources of cheap, easily-transportable arms for desperate customers in areas of violent conflict willing to pay much higher prices
- reliant on personal contacts and networks more than corporate identities
- thriving on corruptible officials and weak law enforcement
- tempted, in some cases, to use fake documentation and bribery which could lead to involvement with smuggling and organized crime[25]

The most celebrated and, at the same time, notorious case of covert government arms transfer was the US CIA's Iran–Contra scandal, which had its origins in the middle of Ronald Reagan's presidency.

Shortly after Reagan's 1984 landslide re-election, members of his administration and others embarked on a series of covert operations. The scheme involved the illegal diversion of about $40 million—acquired from weapons sales to Iran and solicitations to foreign powers and private American citizens—to the Nicaraguan Contra rebels over a two-year period.[26] Two of Reagan's national security advisors, Robert McFarlane and John Poindexter, besides National Security Council staff member Colonel Oliver North, played pivotal operational roles. Because of the scope and scale of various operations and support services, the covert project involved a number of others, including the director of Central Intelligence, William Casey; other senior CIA officers and agents; Department of Defence and State Department officials; private American citizens who served as brokers and fundraisers; foreign intelligence operatives (Israeli and Danish, among others); and foreign nationals who served as intermediaries. An important component of the clandestine weapons transfers to the Contras involved the use of front proprietary aviation companies by the CIA to fly the cargos to the rebels' strongholds in Honduras and Nicaragua.[27]

Illegal arms shipments during the Cold War also fuelled the civil wars in Afghanistan, Angola, Cambodia, Sri Lanka, Ethiopia, Mozambique and some of the Central American countries, where there was a steady demand. Most of these conflicts involved insurgent or terrorist groups, which were not able to obtain arms in the open market. In some of these cases, the US government itself (acting through the CIA) resorted to the black market in order to obtain arms for insurgent groups that it wished to aid covertly. One of these groups was Jonas Savimbi's União Nacional para a Independência Total de Angola (National Union for the Total Independence of Angola) or UNITA, to which the CIA supplied weapons in the late seventies and eighties, using Hercules transport aircraft flying for a mysterious St Lucia Airways. This was later discovered to be the US intelligence agency's air branch proprietary of choice for the delivery of Iran-bound missiles during the Iran–Contra affair.[28] One study concluded that in the decade of the eighties, there were twenty-two reported cases of illegal shipments of small arms from the US to end-users in Iran and the Eastern Bloc. 'These cases included only those commodities actually seized. The value of goods that would have been exported illegally[...]would have been significantly higher,' the study reported.[29]

During and after the end of the Cold War, Latin American countries, including Colombia and El Salvador, were inundated with massive quantities of SALWs, fuelling political conflicts and criminal violence. This is true of even some of the most protracted and bloody conflicts, such as the guerrilla wars in Colombia, Guatemala, Nicaragua and Peru. While reports of the Pakistani ISI's clandestine arms supplies to Kashmiri and other Islamist terrorists in India and the Chinese supplies to the Myanmar military junta and insurgent groups are legion, the post-Cold War period saw SALWs flooding Rwanda, Sudan, Somalia and Bosnia-Herzegovina. In the 1990 civil war between the Tutsi and Hutu groups in Rwanda, 'arms suppliers [in Western and Eastern Europe] rushed to both sides like vultures to a carcass... More than a dozen nations helped fuel the Rwandan war, and both sides appear to have purchased considerable weaponry through private sources on the open market. Former Warsaw Pact countries appear to have supplied both sides, seeing opportunity in Rwanda... Russians, Romanians, Bulgarians, Czechs, Slovaks, and others are aggressively promoting arms sales.'[30] In the nineties, governments and government agencies were the principal suppliers or facilitators in the 'spread of small arms in the overwhelming majority of cases across the world'.[31]

By 1995, analysts had begun to explain the widespread and unchecked transfer of small arms with the 'diffusion/global violence proliferation' model. According to this, 'the key post-Cold War arms trade characteristic was that arms were "diffused" not only to governments, but also to private armies and militias, insurgent groups, criminal organisations and non-state actors'.[32] From a micro-economics perspective, the Purulia weapons consignment 'fulfilled' both the principles of the supply-side push and the demand-side economics. It was just that Indian investigating agencies found it difficult to pinpoint whether an individual or an intelligence agency or an insurgent group paid for the arms. It is yet to be ascertained whether any 'effective demand'[33] articulated itself into a force strong enough to affect the supply and delivery of such a huge consignment of weapons.

Global arms flows often involved a dubious mix of legal and illegal operations. Sometimes, arms dealers, especially in Europe and the US, 'copied the techniques'[34] applied in covert operations by intelligence agencies in camouflaging the purchases and transfers to clients across the world. The networks that controlled and generated illicit arms supplies frequently took recourse to legal arms purchases or transfers, which were subsequently diverted to unauthorized recipients. In 1999, signalling the linkages between legal trade and illicit transfers, the UN Panel of Governmental Experts on Small Arms clearly said:

> Arms brokers play a key role in such networks, along with disreputable transportation and finance companies. Illicit arms trafficking can sometimes be helped by negligent or corrupt governmental officials and by inadequate border and customs controls. Smuggling of illicit arms by criminals, drug traffickers, terrorists, mercenaries or insurgent groups is also an important factor. Efforts to combat illicit arms trafficking are in some cases hampered by inadequate national systems to control stocks and transfers of arms, shortcomings or differences in the legislation and enforcement mechanisms between the States involved, and a lack of information exchange and cooperation at the national, regional and international levels.[35]

Weak Regulatory Regimes

In 1995 and in the early years of the twenty-first century, SALWs manufactured in Western and Eastern Europe and their trans-shipment across continents were not governed by any robust regional or international regulatory regimes. In fact, even as the Purulia arms drop went underway, the UN General Assembly adopted Resolution 50/70B of 12 December 1995, moved by Japan, which went only so far as to request the secretary general to prepare a report with the assistance of a panel of government experts on the question of SALWs in all its aspects.[36] Despite claims to take steps to enhance international cooperation in preventing, combating and eradicating the illicit brokering in small arms, norm-building to control the proliferation of SALWs was excruciatingly slow, especially with little cooperation from UN member states which were either involved in large arms transfers or were recipients of weapons supplies, both licit and illicit.

That little could be achieved to stem the tide of the global arms flow was evident twelve years after the Purulia arms drop, in a General Assembly report in which the secretary general admitted that

> Illicit brokering in small arms and light weapons continues to fuel the illicit trade in those weapons and is largely responsible for violations of arms embargoes imposed by the United Nations Security Council. Those activities facilitated the flow of illicit small arms and light weapons into conflict areas and into the hands of criminal and terrorist groups, with grave consequences for international peace and security, economic and social development and the safety of civilians.[37]

The Purulia arms drop and the numerous players involved in the conspiracy did not figure in any UN document or the agenda of Western non-government organizations (NGOs) that aimed to give teeth to some of the existing fledgling norms and weak regulatory regimes to check the pervasive flow of SALWs, especially to Third World countries, mainly in South Asia where the 'burgeoning trade in SALWs, mostly illicit, has spawned more than 250 militant and insurgency movements'.[38]

On a normative plane, governments proclaimed setting up appropriate national policies and programmes which would be integral, if not prerequisites, to effective action against illicit small arms trafficking. In reality, the 'analysis and policy solutions do not fully address the real world of international small-arms dealers and the method used to carry out arms transfers on the fringes of the law.' In fact, 'most governments are insufficiently transparent about exports of small arms, light weapons and paramilitary equipment, making it difficult for independent researchers to uncover and reliably document the hidden deals and complex routes used by arms fixers.'[39] According to Wood and Peleman, some governments acknowledged the need to take proactive measures 'to provide a more effective regulatory system, but the majority still hesitate to rein in the arms brokers and their shipping agents.'[40] By 2007, when the UN General Assembly was trying to institutionalize regulations to prevent, combat and eradicate illegal brokering in SALWs, 'approximately 40 states' had enacted national regulations to control arms-brokering transactions.[41]

Like many other countries, including those in Western Europe, India was wholly unprepared to deal with the legal issues surrounding the illicit transfer of the Purulia weapons consignment. Purulia, on the western fringes of the Indian state of West Bengal, bordering Jharkhand's Ranchi and Dhanbad districts, was one of the unlikeliest places in India where Kim Davy and the British arms dealer Peter Bleach would airdrop nearly five tons of small arms, including rocket launchers, rocket-propelled anti-tank grenades, 250,000 rounds of ammunition and other assorted weapons. For India, which in the early and mid-nineties had begun to experience the violent convulsions of terrorism in Jammu and Kashmir and widespread insurgency movements in the Northeast, this was the first-of-its-kind clandestine arms drop with the aid of an aeroplane.

There were occasions in the past when organized crime syndicates based in India's commercial capital, Mumbai, had been able to smuggle in a few sophisticated small arms and explosives via the Gujarat coast in the aftermath of the destruction of the Babri mosque on 6 December 1992. In the late eighties, Pakistan provided weapons, including AK-47 assault rifles, to Khalistani terrorists. But the Purulia consignment far exceeded any previous illegal shipments to India, both in terms of quantity and the manner in which the weapons were procured and then airdropped.

At the time, the only legal means available to Indian law enforcement authorities was to invoke the archaic Arms Act of 1959, the Arms Rules of 1962 and the Customs Act against traffickers and 'anti-nationals'[42] (groups or individuals) who faced arrest and other legal proceedings. Beyond these, the Indian authorities had little specialized knowledge of the world in which international arms dealers, brokers and shipping agents functioned. At the same time, there was a lack of any coherent legal regime that could make provisions for preventive deterrent initiatives, law enforcement and punitive actions in all areas of illegal trafficking. These were some of the major factors that impeded a comprehensive investigation of the Purulia arms drop case. Although it was one of the worst sufferers of the SALW plague in the region, India made little effort to contribute to any UN-sponsored arms trade treaty. It was not until 2006 that India began taking bilateral initiatives with some countries to institutionalize arrangements which would help combat the illicit trade in SALWs. The agreements were aimed at 'enhancing cooperation in combating organised crime and international terrorism' and to 'provide for exchange of documentation, information and experience on the activities of persons involved in organised crime and terrorism, including, inter-alia, illicit trade in arms, and funding of international terrorism.'[43] However, all it offered in the face of the international community's drive towards a comprehensive arms trade treaty were platitudes and 'state-centric' and realist positions 'devoid of any concern for public safety'.[44]

What was surprising was that since the nineties and well into the 2000s, India remained one of the largest importers of SALWs from Bulgarian weapons manufacturing units[45] like Arsenal in Kazanlak and VMZ in Sopot, the very factories where the weapons for the Purulia arms drop were produced and subsequently dispatched. While there is no reliable data on the volume of illegal transfer of SALWs to India, save data culled from seizures by the security forces, by 2011 it had achieved the dubious distinction of the 'world's biggest importer of arms (including SALWs), displacing China by accounting for 10 per cent of global arms sales volumes'[46]. According to one analyst, 'the first discordant note had come from India in 2006 when it abstained from voting on [UN] Resolution 61/89 along with a motley group of 24 nations when 153 nations had voted for the resolution.'[47] However, what appeared to have caused India to abstain from voting on the resolution was the 'non-inclusion of non-state actors as likely recipients of illicit arms which, in the Indian experience, is crucial'.[48] India's international commitment was in opposition to its realist stand.

Regardless of India's stand on either conventional weapons or the illicit transfer of small arms in recent years, back in the nineties, the Purulia arms drop case, instead of illuminating the dark world of shady arms dealers and profit-hungry brokers, was allowed to remain hidden in plain sight, both nationally and internationally.

4

'THE INDIANS COULD BE ADVISED
OF THE SITUATION'

Three days after the Copenhagen meeting, on 21 August 1995, Peter Bleach[1] walked into the Northumberland Avenue office building of the Defence Export Service Organisation (DESO)—a department of the UK Ministry of Defence which promotes British defence exports and offers assistance to UK-based defence traders—where he spoke to a certain Stuart Mills, an employee of the UK Ministry of Defence. At this time, Mills occupied DESO's Bangladesh desk and Bleach had come into contact with him during previous business dealings with the Bangladesh government.[2] During this brief but important meeting, Bleach informed Mills about the Copenhagen meeting and sought advice from the DESO employee on how to proceed. Mills referred him to Colin Allkins, who occupied the Indian desk at DESO.

Bleach called up Allkins and explained the situation to him in detail. He asked Allkins if he could put him in touch either with someone at the Indian High Commission in London or, alternatively, with the relevant British authorities. Allkins said that he would inform the necessary British authorities and, in the meantime, it was agreed that Bleach would maintain contact with Haestrup and his business associates so that he could gather as much information as possible. Allkins assured Bleach that someone would contact him very soon to take down the details and give him necessary advice and instructions.

As a matter of abundant precaution and with an eye on the future, Bleach taped his conversation with Allkins, together with several others. He feared that as in some cases UK government departments had refused to release documents pertaining to defence-related cases, which showed that the traders concerned had acted with full government approval, he too might find himself in a fix. The other reason for taping the conversation with Allkins was to safeguard his company's interests. He sensed that he was on dangerous ground. Little did Bleach know, or suspect, that all his telephone conversations were being intercepted by MI5 as well. By the early 1990s, the Security Service, which for years had focused on counter-espionage, had moulded itself to concentrate on counter-terrorism and 'support for the police against serious organised crime'[3]. In fact, the 1989 Security Service Act was amended in 1996 to allow the Security Service to act in support of the police in serious crime investigations. Note how, in comparison, the Indian IB, which has always had the mandate to support the police forces in the states, failed to share R&AW's intelligence with the West Bengal and Bihar governments.

Bleach followed up his conversation with Allkins that very day with a two-page fax letter dated 18 August 1995, written on Aeroserve UK letterhead, in which he outlined in brief the details of the Copenhagen meeting and the 'quote for sale of AK-47 rifles and ammunition'[4]. The letter continued:

> Earlier this week, I was informed that my quote was acceptable and I was invited to discuss the final details with the buyers who were acting on behalf of the end-user. This turns out to be an illicit deal and the end-user appears, at this stage, to be an insurgent group in India. During the meeting, I was asked for suggestions on final delivery.
>
> Delivery of the goods would require that an aircraft file a commercial flight plan through Indian airspace and passing directly over the area in question. The aircraft would then make an unscheduled landing on a rough airstrip in an area controlled by the insurgents, unload its cargo, and continue on its way. It would probably [sic] on the ground for less than five minutes and it [sic] most unlikely that this activity would be noticed by the Indian authorities. A Russian-built

aircraft requiring only a very short landing strip would be used.

I have not yet declined to supply these goods. I told the buyers that I would need more time to calculate the additional costs of a clandestine delivery and that I would respond in a couple of days. The buyers are very anxious to conclude this deal.

It seems to me that there are three options.

Firstly, I could simply tell the buyers that I cannot help them. If I do this, then I am not involved in any illegal activity and I can get on with life. However, if I do this, then they will simply go elsewhere and somebody else will provide the goods.

Secondly, I could either simply supply the goods FOB [free on board] and give them the name of a Ukrainian or Moldovan freight company who will land anywhere for money; or of course the deal could be completed in its entirety. (This is not put forward as a serious suggestion).

Thirdly, the Indians could be advised of the situation and the deal could be allowed to run. If the delivery is to be made into the heart of the insurgent area, then it seems to me that it may be of interest to establish precisely where this location is. The shipment is certain to be met by a very large number of insurgent personnel, including some of their leaders. The problem could no doubt be dealt with by the Indian Air Force whilst the delivery was taking place.

Advice is currently required and until I receive it I will continue to communicate with the buyers as normal. The buyers, incidentally, are of Western European, non-British, origin and certainly involved in other illicit deals, some details of we [sic] I have.

A 10 June 1998 letter of the Organised and International Crime Directorate (OICD) of the UK Home Office, addressed to Simon Scadden Esq, then British deputy high commissioner in Calcutta, is revealing. It says that Bleach's letter to Allkins was 'passed to the

Security Service' or the British MI5.[5] The letter further says that the
North Yorkshire police force was asked, on 22 August 1995, to
interview Bleach to obtain further details of the arms deal. The police
officers conducted the interview on 14 September, following Bleach's
return to his address in their area. During the interview, Bleach told
the police officers that he was then still seeking advice on whether to
proceed with the arms deal.[6] We shall return to the 14 September
1995 meeting between Bleach and the officers of the North Yorkshire
police shortly, but before doing so it will be worthwhile to know what
Bleach did after he had informed Allkins, who most certainly shared
the contents of Bleach's fax to him with the MI5, of what had
transpired.

Between 27 August and 7 September, Bleach was in Dhaka,
attending to outstanding contractual business with the Bangladesh
Department of Defence Procurement. A performance guarantee
required for one contract had been lodged in an incorrect format.
That necessitated Bleach's presence in Dhaka where the matter would
be sorted out. He also held meetings on the subject of delivery of two
helicopters for the Bangladesh Ministry of Relief.

Bleach's Dhaka Sojourn

Bleach, who put up at the Contessa Guest House in Dhaka's posh and
upmarket Banani district, pondered over the details of the Copenhagen
meeting. He had no particular knowledge of the cost of an illegal arms
delivery and had no real way to find out. 'So, I made it up,' Bleach
wrote in his statement to the CBI.[7] He informed Haestrup that the
unit price would remain the same, but he would require a cash advance
payment of $50,000, which would be non-refundable under any
circumstances, and that delivery by air would be $70,000 with an
additional negotiable cash bonus for the crew. He arrived at the sum
of $70,000 by finding out the approximate cost of leasing an aircraft
for a ferry flight from Central Europe to Calcutta and simply doubling
it. He gave Haestrup this information by telephone before leaving
Bangladesh.

Soon, Bleach received from Brian Thune a fax copy of a letter of
credit for the sum of $460,000. Although Thune claimed that the

letter of credit had already been issued, Bleach found out later that it was issued by Kim Davy only at Thune's request. The letter of credit was signed by Kim P. Davy, which at that time meant nothing to Bleach. To Bleach's surprise, the terms of the letter of credit bore no relation whatsoever to anything that had been discussed. It was poorly written and he noted that the beneficiary was Aeroserve UK/Danacot Steel (Thune's company). Bleach later mulled over this matter and concluded that Thune had tried to steal this money from Davy, but at the time, he found the matter annoying. He returned the letter of credit to the issuing bank (Hang Seng, Hong Kong).

'DNA'

Davy, alias Nielsen, was not sitting idle either. On 26 August 1995, an email (which most likely originated from Hong Kong) signed DNA[8], which was his email user ID, and addressed to a certain Ramesh[9], said:

> I am going to come complete straight [sic] with you but you must promise me that the details of this remains between us, you can off [sic] course let Shyam [possibly another Ananda Marga member] know the general things but not [the] details. In the last three-four months I have lost several hundred thousand dollars just as I was getting confident having started several business things and investing money here and there to make sure that nothing could blow us out of the water all the way [sic]. Well, the Lord has His ways, somehow one after the other money was stolen, people were in trouble etc. I lost it all within a few months. Then I stopped communicating because I did not know what was going to happen.

In the message, DNA also asked Ramesh, 'What about alternative ways of getting funds for the project?... My crisis has also meant that there has been tremendous crisis in the AN about the construction there.'[10]

A week after that (2 September 1995), Nielsen left Hong Kong for Johannesburg, but five days later he was apparently back in Hong

Kong since a fax saying 'Hong Kong 7 September, 1995' was prepared for a certain Ivan Whitehead living in Johannesburg. The fax's contents read, 'Decision makers coming to SA 23/24 September. The technical team will come on 15ᵗʰ September.' [11] It was signed 'Peter Johnson'. When the South African police, on the request of Interpol officials, executed a check on Whitehead, it was found that such a person, born in 1957, did indeed exist but had no criminal antecedents. Apart from the information that Peter Johnson was another of Nielsen's many assumed names, the check did not turn up anything linked to the arms drop conspiracy. Nielsen left Hong Kong for Denmark on 8 September 1995, a week before finalizing the contract for the weapons with Bleach in London.

The Bangkok Plan

Back in North Yorkshire on 7 September 1995, Bleach found, to his amazement, several messages on his answering machine from Haestrup, pleading with him to call urgently. When Bleach immediately returned the call, Haestrup said that Bleach's proposals had been broadly accepted, but in order to get final approval, he would have to fly to Bangkok for a meeting with 'the Boss'[12]. Yet again Bleach told Haestrup that he would be happy to fly to Bangkok provided that the airfare and other expenses were paid for. Haestrup said this would be arranged and advised Bleach to reach the Thai capital on the evening of Wednesday, 27 September 1995.

When Haestrup said that he wanted to fly with Bleach so he could discuss business before the meeting, the Briton agreed, but said that he had to make a stopover in Dhaka ahead of the Bangkok meeting. Bleach suggested the two meet in Dubai, which was an excellent hub for destinations to Southeast Asia and the subcontinent, while being easily accessible from Europe. Haestrup agreed.

Bleach called his travel agent and asked him to book a ticket for Bangkok for 24 September 1995, with stopovers in Dubai and Dhaka, which would reach the Thai capital on 27 September. He advised the travel agent that he wanted to fly Emirates and that from Dubai onwards the ticket should be open. A few hours later, the travel agent called Bleach to confirm the purchase of the tickets—Emirates till

Dhaka and then Royal Thai Airways from Dhaka to Bangkok. Bleach's return to the UK from Bangkok was again booked on Emirates via Dubai. The cost of the round trip was £980.

As soon as the bookings were made, Bleach called Haestrup and provided him the details of his itinerary. He told the Dane that when in Dubai he always stayed at the Al Khaleej Palace Hotel, near the old town docks, and suggested that the two meet sometime on 24 September, to which Haestrup agreed. On the night of 21 September, Haestrup called Bleach at his Yorkshire office and requested him to purchase tickets on Emirates, since he found the airfare from Denmark to Dubai and onwards to Dhaka and Bangkok rather costly. Bleach promised to buy the tickets through his travel agent and the two agreed to meet at Manchester Airport around 1100 hours GMT on the morning of 24 September.

The Deal Is Inked

Two weeks prior to their departure, several important developments took place. On 13 September, Bleach sent a message addressed to his agent in Bangladesh, Captain (retired) Syed Tasleem Hussain, informing him that he had 'faxed a copy of the End-User Certificate to our suppliers who have said that in its present state [it] is not acceptable'[13]. Bleach then dictated to Hussain the wording of a new EUC, an indication that Bleach had strong and reliable contacts in Bangladesh since he was able to order a 'specially made' one.

The same day, Bleach met Nielsen at the Intercontinental Hotel in London, where a contract for the sale of goods, i.e. arms and ammunition, was signed by the two (Nielsen signed in the name of Kim P. Davy) on Aeroserve UK's letterhead. Nielsen was accompanied by his Britain-based lawyer, who did not identify himself at the time. As the 'buyer', the contract document showed Davy's company as Howerstoke Trading Ltd, 89, Queensway, Hong Kong. The description of the goods—2,500 AK-47 assault rifles, Type 56/2; 7.62 calibre with folding metal butt and plastic furniture; each rifle supplied with two magazines and cleaning kit. The unit price was marked at $98 and the total FOB price was $245,000. The cost of 1,500,000 rounds of ammunition was marked at $135,000. The total air freight and handling charges were $70,000. Barring the handling charge, which was to be payable in full and in advance, the rest of the cost ($415,000) was payable by letter of credit.

As a note, the contract said that a valid EUC would be required after the contract was signed, subject to signing of a non-circumvention agreement, which too was inked that very day. The deal was struck and it was only a matter of time and a few more details, including the purchase of a viable aircraft and hiring the crew, before the operation to airdrop the arms and ammunition over Purulia could be underway. Everything appeared to be running like clockwork. What Bleach withheld from his CBI interrogators and the British police was that the 'package deal' with Nielsen involved imparting weapons training to members of the insurgent group over a period of six months, during which time Bleach would visit Purulia every two months to review the progress made.

On 13 September 1995, Nielsen used his Citibank Visa card for the withdrawal of an unknown, though not a big, amount at Hounslow, near Heathrow. This was another 'footprint' that Davy left behind for Indian and Interpol investigators to follow months later.

Cops Come Calling

A day after the deal was signed, two officers from the North Yorkshire police, Special Branch, knocked on the door of Bleach's Howdale Farm residence. There is evidence to show that the British Home Office was aware that the North Yorkshire police force was asked, on 22 August 1995, to interview Bleach to obtain further details of the arms deal. After two failed attempts to meet Bleach on 29 August 1995 and 30 August 1995,[14] the two policemen finally met him at 1150 hours on 14 September. They had come armed with seven broad questions that the MI5 had instructed one of them, Detective Sergeant Stephen Leslie Elcock, to put to Bleach. The other policeman, Detective Constable Jones, noted down Bleach's replies to the questions, among which were:

> (a) Can he [Bleach] provide details of the arms and ammunition he has been asked to supply? (b) Will this

equipment be consolidated with arms procured from other dealers, and if so, can he provide details? (c) What is the value of the business and how is payment to be arranged? (d) Details of the approach to him – when, where and who from? How does he contact the buyers and what are their contact details? (e) Details of the insurgent group in India? (f) What area do they wish the arms delivered to and when and where is the rough airstrip?[15]

Like most of its operations at home, the MI5 had named the inquiries with Bleach and its subsequent findings Operation GAINFUL and Operation LEMON.

The police noted from the meeting that 'Advice was given to Bleach in that we would re-contact him to advise him regarding his proposed dealings with his Danish contacts, i.e. Bleach was running until advise'[16]. Subsequently, Detective Superintendent Ian Lynch of the Special Branch said in a statement, 'One [sic] my officers documented the reply from Peter Bleach and notified the Security Services (MI5) so that left no doubt that the deal was continued until Peter Bleach was given appropriate advice. There is no doubt that Bleach was advised.'[17]

Curiously enough, and contrary to Detective Superintendent Ian Lynch's version, Sergeant Elcock said in his statement that at '0915 hours GMT on Friday, September 22, 1995, as a result of further communication, I again visited Howdale Farm with Detective Constable Jones where I again saw Bleach'[18]. Sergeant Elcock told Bleach that 'there was advice in relation to his dealings' and that he pointed out to Bleach in 'clear terms' that he should 'withdraw from the proposed deal with his Danish counterparts in relation to the arms deal.' Sergeant Elcock also said that the details of such a deal would be passed on to the Indian authorities. Bleach accepted the advice, saying that it would be no problem, and he would tell his Danish contacts that he was unable to meet the original quote.[19]

During a third meeting on 8 December 1995, according to Sergeant Elcock's witness statement, Bleach told him that 'Thune had disappeared with money belonging to a Hong Kong gold merchant', that Haestrup, Kim Davy and a Wai Hong Mak had purchased an

Antonov-26 aircraft of which he was the broker and that the cost of the aircraft was $250,000. Detective Constable Jones, who took copious notes, recorded in his handwriting that the aircraft would be used to carry freight from Bangladesh to Varanasi. Bleach carried on sharing information, stating that the original arms deal was, in his opinion, used to assess the viability of gold importation into Pakistan and India.[20]

The discussions that took place on each of the three occasions the Special Branch police met with Bleach were reported to the 'relevant' British authorities, in this case, operatives of the MI5. Bleach later claimed, while in the custody of the CBI, that he was given the impression that the Indian authorities would be duly informed and that he was asked whether he would be prepared to 'run' with the case, certainly for the time being and possibly to its final conclusion, so that the various participants could be properly identified. Bleach claimed that he agreed to this quite willingly.

It is abundantly clear that by this time, British intelligence knew much of the details of the clandestine operation and the principal participants. MI5 most certainly took its time to assess the importance and credibility of the information Bleach provided and subsequently shared all of it with the Security Intelligence Service (MI6), the foreign arm of British intelligence. What caused the delay on the part of MI6 to share Bleach's information with Indian intelligence? Was British intelligence, MI5 as well as MI6, trying to ascertain the likely involvement of a security agency of another state power, with which it enjoyed very warm and friendly relations? These are questions to which there are no simple answers. After all, covert intelligence operations, by their very nature, are and can be deniable and little, if any, trace linking them to any particular agency is ever found. My sources in the Indian intelligence community told me that MI6 did indeed share the information in as much detail as possible with the R&AW on three occasions, from 3 November 1995 through early December 1995, by which time Bleach was completely committed to the arms drop operation, claiming to the Special Branch that he 'would be happy to work' with the objective of defeating the purpose of the insurgent recipient group. Confident and happy that the Indian authorities would be kept informed of the developments, Bleach set out for Bangkok for the crucial meeting with Nielsen, Haestrup and a few others who were to play key behind-the-scenes and on-the-ground roles before and on the night of the arms drop.

The Bangkok Meeting

As agreed on 21 September 1995, Bleach and Haestrup met at Manchester airport on 24 September and flew out together, arriving in Dhaka on the morning of 25 September after a stopover in Dubai. At Dhaka's Zia International Airport, Bleach and Haestrup were met by the Briton's agent, Syed Tasleem Hussain, who drove them to the Contessa Guest House, where the duo occupied separate rooms. Bleach spent 25 and 26 September—Monday and Tuesday—making courtesy calls at the Bangladesh Ministry of Defence Procurement and fixing the earlier problem of a faulty performance guarantee. Haestrup did not attend any of these meetings. On both days, the two had lunch together at Dhaka Club where they met a number of local businessmen on a social basis. They had dinner at Hussain's house with his family on 26 September, before flying out to Bangkok on the afternoon of 27 September.

At Bangkok's Don Mueang International Airport, Bleach and Haestrup were met by Nielsen who, for the first time, introduced himself to the British arms dealer as Kim Davy. The Dane, dressed in faded blue jeans and a T-shirt, had a car waiting. He drove his guests to the five-star Felix Hotel located in central Bangkok. Arriving at the hotel around 1930 hours local time, Haestrup and Bleach checked into the rooms which were booked by Nielsen and secured against his credit card. Nielsen told Bleach that a meeting had been arranged at the business centre of the hotel between 2100 hours and 2130 hours, so he had enough time to take a shower and change. Dinner had been arranged at the business centre.

A separate room had been booked exclusively for the meeting where, other than Bleach, five other persons were already present— Peter Haestrup; Niels Christian Nielsen; Nielsen's lawyer, who identified himself as 'Bryan'; an Indian who was introduced to Bleach as 'Randy'; and a Chinese by the name of Wai Hong Mak, a resident of Hong Kong. Nielsen performed the introductions and although, at some stage of the meeting, he did mention what was alleged to be Randy's real name[21], Bleach could not recollect it later while in the CBI's custody. Randy was described as being in charge of the insurgent group in India. He was dressed in a very simple fashion, wearing a smock of an indeterminate rusty colour, wrinkled European-style trousers and no shoes. Everybody showed great deference to Mak, who appeared to have the final say in financial matters. According to Bleach, Mak's business was stated to be in gold dealing and Nielsen was a close but clearly junior partner in this business.[22] Bleach never found out the full name of Nielsen's lawyer Bryan who was white, about 5'10", of medium build, had short, well-trimmed hair, a thin nose and thin lips. He appeared to be of European origin and had a peculiar accent, which sounded like a cross between Australian and South African. A resident of London, Bryan spent much of his time in Bangkok. He was very well dressed in casual but conservative British clothing.

Randy, alias Satyendra Singh, alias Satyanarayana Gowda.

The meeting began with dinner which was ordered by Bryan, who was apparently familiar with Thai cuisine. Bleach noted that both Nielsen and Randy were strict vegetarians. When Bleach commented on this, Nielsen quickly pointed out that Randy and he also did not eat onions or garlic; they did not consume alcohol or tobacco, and only drank fruit juice. If fruit juice was not available, they drank water.

After dinner, Nielsen produced two survey maps of India, both with 'West Bengal' marked, a larger map covering a very large portion of eastern and central India, and a number of photographs. As the men in the room huddled over the table, Nielsen spread out the larger map, indicating to Bleach the general area where the arms would be delivered. This was vague enough for the Briton to avoid precise identification, but more than enough to permit an accurate calculation of costs and other logistics. Nielsen then turned to the survey maps, which were very detailed and probably on a scale of 1:100,000 or less. He marked what appeared to be two small but prominent hills on an otherwise flat and relatively featureless area. It was clear from the maps that the contour lines were very close together, indicating that the hills were quite steep and that one of them had a ravine down the middle, covered with light forests.

Bleach concluded from the maps that the hills would be very prominent landmarks. Besides, he saw at least one marked house. Nielsen put his finger on the house and said that it was occupied by 'friends'[23]. He turned towards Bleach and in a measured tone told him that this was the precise area where the goods were to be delivered and that if it was necessary to land a plane, Randy and his men would have no difficulty in clearing the place to make for a landing strip. However, he then said that Randy and his men would prefer to have the goods dropped by parachutes, as that would leave less evidence of the delivery. Someone suggested that the arms and ammunition could be dropped into the ravine on the larger of the two hills, so that they would fall into the jungle. Nielsen assured Bleach that Randy had more than sufficient men available to collect all the goods and conceal them in a very short period of time. He also said that Randy had in his command several jeeps, which could be used to move the weapons.

Nielsen then showed Bleach the photographs, which appeared to have been printed from slides. They showed the general view of the area that was shown on the survey maps and he said that he had taken them himself since they last met in Copenhagen, slightly over a month ago. The area immediately around the hills appeared to be rural and thinly populated. Bleach told Nielsen that the proposed site looked ideal to him. Nielsen, completely focused, informed him that Randy and his men had absolute control over an area of 20 square kilometres from the hills and that no security forces ever ventured into that area. He was confident that the weapons and ammunition would be collected in complete safety.

Throughout the lengthy conversation, neither Mak nor Randy spoke a word. Nielsen did all the talking.

When the Dane asked the Briton about the aircraft, Bleach said that a Russian medium transport plane would be ideal and there would be no problem in leasing one at a price he had earlier quoted. Bleach had carried a copy of *Jane's All the World's Aircraft*, which provided comprehensive and detailed reference information on the production of nearly all civil and military aircraft in service across the globe at the time. He quickly leafed through the manual to show the assembled group photographs of several models.

Clearly satisfied with the proceedings so far, Nielsen then brought up the subject of gold dealing. Facing Bleach, he told the Briton that he and some of his associates regularly delivered gold to clients across Asia and the Indian subcontinent, and he wondered if this could be arranged by air. It was not immediately obvious whether he was talking about legal or illicit trading, though it later transpired that Nielsen was hinting at a combination of the two. When Bleach asked him how often deliveries were to be made, Nielsen said every two weeks.

Bleach cautioned that while there would be no problem in leasing an aircraft every fortnight, the gold-dealing operation would not be economically viable. The cost structure meant that within six months the lease costs would be far greater than the actual cost of purchase. He suggested that Nielsen consider purchasing a second-hand Russian cargo plane, which could cost an estimated $200,000. Besides, spares were inexpensive and as the type was common in the area, they would be easy to service and maintain. A full crew, including ground maintenance men, could be hired from Russia at a very reasonable price.

Bleach now advanced sound business ideas. He pointed out that there was a great shortage of freight aircraft in the area (the Indian subcontinent) and suggested that in between delivering gold, the aircraft could make a very healthy living out of delivering freight to large airports across the subcontinent. The net effect would be that the aircraft would probably pay for itself in about a year and would begin making profits in the second year of its operation. Nielsen discussed this with Mak for a while before asking Bleach for his suggestions regarding a suitable aircraft. Bleach suggested that an Antonov AN-26 would be ideal and produced a picture from *Jane's All the World's Aircraft*, pointing out that its Western equivalent, a Fokker, was several times the cost and not as efficient. A technical discussion followed on the various merits and faults of an AN-26. Finally, Mak said that he would consider it overnight and come up with an answer in the morning.

On his part, when Nielsen asked Bleach whether the same aircraft would be suitable for dropping the arms by parachute over West Bengal, the arms dealer said that he did not see any reason why not and that it could probably make a landing as well, since it was designed to operate on unprepared airstrips. Nielsen had another question: where would the aircraft be based? Bleach said he thought Bangladesh would be a good idea, since it was an underdeveloped country which offered great business opportunities. He volunteered Syed Tasleem Hussain's name because Hussain was well-connected in business circles and could probably introduce Nielsen and his associates to the right people or companies for forming a joint venture.

Nielsen reverted to the discussion over the weapons. He said that he had never wished to purchase 2,500 guns, because such a large number would be inconvenient for storage. He had only agreed to that number because Brian Thune has said that it was the smallest order available. Nielsen told Bleach that Thune had given a deposit of $150,000 against the delivery of the guns and that he had since disappeared with the money. The money had not been recovered and Nielsen and his friends were taking steps to track down Thune, recover the cash and punish him for his actions. Bleach gained the impression that such punishment would be very severe indeed.[24] From a more realistic perspective, Nielsen said that it would be more convenient for him to purchase only 500 guns and the corresponding suitable quantity of ammunition. To this, Bleach said there would be no problem. So Nielsen breathed easy, offering to purchase the rifles and the ammunition, together with the aircraft, for no more than the original cost of the 2,500 rifles. When Bleach agreed readily, Nielsen asked him to prepare a new costing, which he assured would be approved.

As the meeting broke up, Nielsen, Randy and Mak retired to their rooms. Bleach and Haestrup went downstairs for a drink at the hotel bar along with Bryan, who in turn was joined shortly by his girlfriend, a well-dressed Thai woman in her early thirties. The other men were introduced to her and had two or three drinks together before the group split up for the night. No business was discussed, but Haestrup took the opportunity to tell Bleach that when Randy arrived in Bangkok he was not wearing shoes, and a pair of sandals had to be urgently procured for him. Haestrup also told him that Randy and Nielsen shared the same room in the hotel.

The next morning at breakfast (28 September 1995), Mak joined Bleach briefly and told him that he had approved the aircraft deal, and that he could sort out the details with Nielsen. Mak left immediately thereafter and took an early flight to Hong Kong. Bleach met Nielsen and Haestrup, with whom there was a brief discussion over the aircraft. It was agreed that Bleach should immediately leave for Europe to try to locate a suitable plane. In the meantime, Nielsen and Haestrup would stopover in Dhaka, meet with Syed Tasleem Hussain to discuss the joint venture company and get the necessary administrative and other paperwork done. Haestrup would visit Dhaka several times in the next three months to desperately procure landing permits for the AN-26, even trying to bribe the top bosses at the Civil Aviation Authority of Bangladesh.

Bleach told Nielsen that he would require $35,000 as an advance deposit against expenses for the aircraft procurement. He explained that the process would incur heavy expenditure which he could not afford on his own. Besides, he was not prepared to take the risk that Nielsen and his associates might change their mind halfway through. Nielsen did not raise a fuss, saying that he would organize the payment immediately. Bleach gave Nielsen his bank details. At that point, Nielsen let out a secret—Randy had arrived the previous day from Calcutta via Dhaka on a Biman Bangladesh flight, for which he had a fixed-return date ticket. Randy, according to Nielsen, had expected the meeting to last a few days and had not anticipated returning the following day. Since the next flight to Calcutta that day was at 1100 hours Thailand time, Nielsen requested Bleach and Haestrup to take Randy down to the Biman Bangladesh office in Bangkok and try to get his ticket changed so that he could return the same day.

The three—Haestrup, Bleach, and Randy—took a cab to the Biman office, where they found that it was practically impossible to change Randy's ticket, which was also non-refundable. The only alternative was to purchase a new ticket. Biman's flights were fully booked for the day, so the lady behind the counter recommended a nearby travel agent to purchase a ticket on another airline. At the travel agent's office, they found that Randy could get a seat on either an Air India or an Indian Airlines flight, which flew direct from Bangkok to Calcutta and would depart in the early afternoon. Randy purchased the ticket and, to do so, produced an Indian passport. Time was short, so Randy was put into a cab, with the driver told to take him to the airport as fast as possible through Bangkok's crowded and congested streets. Randy was advised that if he missed the flight, he should take a taxi back to Felix Hotel and wait for Bleach and Haestrup there. The cab sped away and Bleach never saw Randy again.

Haestrup and Bleach then went shopping and had a late lunch before returning to the hotel in time for the Dane to check out at around 1700 hours to take a flight to Hong Kong. Bleach's own flight would not depart until 0100 hours the following morning. So he took a shower, changed into fresh clothes and checked out of his room at around 2130 hours. All bills were secured by Nielsen's credit card, so Bleach did not have to pay for anything. He moved to the bar and had a couple of drinks before hiring a cab to reach Don Mueang Airport around 2300 hours. The Emirates flight to the UK was on schedule, with a stopover in Dubai, where Bleach again put up at Al Khaleej Palace Hotel. He arrived at Manchester on Saturday, 30 September 1995. To his dismay he found out that Haestrup had not refunded the money for Bleach's flight ticket and, in fact, had borrowed £500 from Bleach to purchase his own onward ticket to Hong Kong. Putting that thought away for the time being, Bleach got on with more urgent and pressing work. So did Niels Christian Nielsen.

5

'THE RIGHT AIRCRAFT HAS BEEN FOUND'

Before reaching Bangkok, Nielsen used his Citibank Visa credit card at least four more times, leaving behind tell-tale signs of his travel from the UK (after signing the contract with Bleach on 13 September 1995) via Denmark, the Netherlands and finally India.

- On 15 September 1995, two money withdrawals were recorded— one in Frederiksburg, approximately 60 kilometres west of Copenhagen, and the other at Kastrup Airport in Copenhagen, Denmark.
- On 16 September 1995, a third transaction was recorded in the Netherlands at Schiphol Airport.
- On 17 September 1995, a money transaction was addressed to the Caransa Hotel in Amsterdam, Holland. Nielsen apparently checked out of the hotel on his way to take a flight to Calcutta.[1]

The Mysterious William Johnson

A day later, Nielsen checked in at the Great Eastern Hotel in the heart of Calcutta's commercial district, not far from the governor's house or Raj Bhavan, the West Bengal government state secretariat and the city police headquarters. Along with Nielsen, another man, a certain William Johnson, also checked into the hotel. On 19 September, Nielsen took a train from Howrah station to Purulia, where he de-trained at Pundag station, scaled the perimeter wall to avoid being noticed by policemen and headed for Ananda Nagar, the global headquarters of the secretive, right-wing cult organization, the Ananda Marga. He met up with Randy and a few other members of the 'insurgent group' before proceeding to a hilly area in the region to take photographs of the precise location where the weapons were intended to be airdropped by an aeroplane.

The bills signed by Nielsen, in the name of Kim P. Davy, were solid evidence of his visit to Calcutta, though there was no indication but for the checkout date from the old hotel's register that he left the city within less than a week. Months later, investigations by the CBI and the IB would find that Nielsen met top Ananda Marga bosses (who were not named) and visited Purulia to take pictures of the probable arms drop site. The real identity of Johnson, who had visited Delhi along with Nielsen earlier in December 1994 and January 1995[2], could never be established. The name Johnson was suspected to be an alias used by a person 'representing a British authority'[3]. According to preliminary findings of the Indian investigation, the same person was suspected to have been at Burgas Airport, Bulgaria, on 10 December 1995, when the AN-26 aircraft was loaded with the arms and ammunition. Johnson's suspected presence at Burgas Airport was never pursued, although Brian Thune had revealed during interrogation in October 1996 that he was the one 'who would sanction the arms covered by the deal'[4]. Johnson was also suspected to have visited Calcutta in July 1995, as an entry in the pocket diary of Thune indicates.[5] The entry, recorded at 0800 hours on 17 July 1995, mentioned the name of an Inspector William Johnson, Calcutta. All that Indian intelligence learned about Johnson was that he held a British passport (No. 017399657) issued in Glasgow on 23 March 1995. Further inquiries revealed a William Johnson, born on 12 June 1962, in Sunderland, who had died on 10 January 1969. It is very likely that the fake identity of William Johnson was used by an American national.

Both the CBI and the Interpol spent months trying to find out 'Johnson's' real identity. They collaborated with British intelligence, the R&AW, and other Western law enforcement agencies, but nothing came up to even remotely indicate his background or antecedents. Was he a British national? Was he an American? He seemed to have vanished into thin air. It was obvious for CBI officers that Johnson was an assumed name to conceal the real identity of the man. But they were loath to admit that Johnson could have been an intelligence operative of a foreign agency. This was, after all, not the line of investigation that the CBI wanted to pursue, for it did not have the capacity or the mandate to launch a counter-intelligence operation. There were two reasons for this. First, the CBI top brass had decided that it would be worthwhile to focus the attention of the probe on the facts and circumstances leading to the arms drop. Second, since they had by then arrested Bleach and all of the AN-26 crew members, it would have been more fruitful to narrow the scope of the investigation in order to make more arrests in the future and eventually mount a solid prosecution case to seek conviction of those accused in the arms drop case.

Meanwhile, Bleach, who was fully aware of Nielsen's secret Calcutta visit, again received a telephone call from Sergeant Stephen Leslie Elcock of the North Yorkshire police, Special Branch, sometime in the second week of October 1995. Bleach gave a full update of the developments, including the Bangkok meeting, and was allegedly assured that he would very soon be informed on his next course of action—whether to completely dissociate himself from the weapons buyers or continue dealing with them.

An Aircraft Is Found

Earlier, on 1 October 1995, Bleach received the $35,000 he had asked of Nielsen, by bank transfer to a Barclays Bank branch on St Nicholas Street, Scarborough. The transfer, from a Kowloon branch of the Hong Kong and Shanghai Banking Corporation (HSBC), had been cleared by Mak, who was apparently responsible for all financial matters. As soon as he received the money, Bleach started the process of locating a suitable aircraft. He telephoned all the available aircraft brokers, including William Roeschke, and had a check done on computerized sales lists. He expected the search job to be long and tedious, but he rapidly learned that two suitable aircraft were available in Riga, Latvia. Apart from those, one was available in the Ukraine and the other in Omsk, Russia. He narrowed his search down to the two aircraft in Riga, which belonged to the former state airline, Latavio Latvian Airlines, that was faced with bankruptcy, because of

which its entire fleet was up for sale. Bleach contacted Geoffrey
Harold Rosenbloom, the managing director of Computaplane, in the
UK. Rosenbloom's company was involved in the buying and selling of
aircraft, helicopters and spares worldwide. On behalf of Bleach,
Rosenbloom located an AN-26 aircraft in Riga. On 3 October 1995,
Rosenbloom sent Bleach a fax message, sharing the news that an
aircraft had indeed been found.[6] During his search for an ideal aircraft,
Bleach had also requested his former commanding officer in British
military intelligence, Colonel John Hughes-Wilson (under whom he
had served in Northern Ireland during the Troubles), who at that
time held a post-retirement job with the North Atlantic Treaty
Organisation (NATO) in Brussels, Belgium. Towards the end of
October 1995, Col Hughes-Wilson telephoned a Jurmala-based
journalist, Mikhail Bruks, whom he had been introduced to by one
Noel Dawes[7] the month before, seeking contacts for his 'friend Peter
Bleach' in Latvia.[8]

On 7 October 1995, Bleach left the UK for Riga, for a preliminary
inspection of the aircraft owned by Latavio. He met Viktor Masagutov,
general director of Latcharter, a Riga-based charter airline company.
Masagutov later became the broker who arranged the sale of the AN-
26 aircraft to Nielsen. Bleach signed a provisional agreement with
Latcharter to temporarily secure the aircraft. A provisional price of
$250,000 was set up, subject to full overhaul, certificate of
airworthiness and a replacement engine.[9] The aircraft, whose call sign
was YL-LDB, appeared to be in good condition. Returning to the
UK three days later, Bleach telephoned Haestrup, informing him
about the aircraft. On 11 October 1995, Nielsen called Bleach,
congratulating him for locating the aircraft quickly, but also seeking
more details about the AN-26. Bleach explained to Nielsen that
although the price was a little high, it could be compensated for by the
fact that it would be much easier and safer to do business in Latvia
than Russia. Nielsen agreed. They decided to meet in Riga on 30
October 1995, to inspect the aircraft together. It was during this
phone conversation that Nielsen let Bleach know that he was headed
for Calcutta for discussions with the insurgent group.[10]

Two days later, Bleach sent a fax message to Masagutov, confirming
the 'intention to purchase the Antonov An-26 aircraft at the agreed

price of $250,000'. Bleach also confirmed that he would like to discuss the 'situation regarding the crew etc' and that he was 'looking for parachute equipment for dropping cargo from an aircraft similar to an AN-26.'[11]

Nieslen had also been busy during this time. Between 3 and 23 October 1995, he travelled frequently to different countries. The following money transactions against his Citibank Visa account were recorded, providing a clear picture of the places that he visited, though not of the persons he met:

- On 13 October 1995, Nielsen left Hong Kong and returned ten days later. On 17 October 1995, a money transaction was made in South Africa, which indicated that he had visited that country.
- On 24 October 1995, Nielsen left Hong Kong for Germany. The next day, a Visa payment was made in Irschenberg, Germany, at a gas station on Highway Number 8 from Munich to Salzburg, about 100 kilometres south of Munich.
- On 26 October 1995, a payment of $111 was made from his Visa account, addressed to Absalon Hotel in Copenhagen, Denmark.
- On 27 October 1995, a Visa payment of $227 was made to Hertz Rent-a-Car in Copenhagen.
- On 29 October 1995, a Visa payment of $252 was addressed to Le Port Restaurant in Espergarde, Denmark. Espergarde is a small village situated about 30 kilometres northeast of Copenhagen.

The Riga Meeting

On 29 October 1995, Bleach travelled to Riga, the Latvian capital. He had checked into Riga Hotel in the city centre. The following morning, Bleach met Masagutov[12] to discuss commission levels in the event that the transaction was successful. Together they went to meet Nielsen, Peter Haestrup and Wai Hong Mak, all three of whom had checked into the Radisson SAS Hotel, a brand new five-star establishment. The three insisted that Bleach check out of Riga Hotel and join them at the plush Radisson Hotel. Bleach agreed and did so in no time.

The Russian Antonov AN-26 aircraft that was used to drop the weapons over Purulia.

In the afternoon, the four inspected the aircraft briefly, following which there was a detailed discussion on the documents that would be required, the plane's general condition and the terms of the purchase. On the morning of 31 October 1995, there was a detailed inspection of two aircraft and the YL-LDB was chosen. After agreeing to the purchase, Mak and Haestrup took the afternoon flight to Copenhagen, en route to Hong Kong. Bleach and Nielsen remained in Riga.

Nielsen had previously made arrangements with another associate called 'Misha' in eastern Russia to have two engineers sent to Riga to inspect the aircraft and give a full report on their condition, if necessary. Shortly before he arrived in Riga on 30 October, Nielsen had called Bleach to give him Misha's (see Chapter 8) telephone number and had advised him to arrange for an inspection. Before reaching Riga, Bleach got in touch with Misha, giving him all necessary details of the aircraft's location, together with Masagutov's phone number by fax. The engineers had arrived in Riga shortly before Bleach and Nielsen and had already commenced work on inspecting the aircraft. They submitted a report on 1 November 1995, recommending an engine replacement for the YL-LDB.

The same day, Bleach and Nielsen went to the offices of Latavio, in central Riga, for a discussion with the airline's chief executives. After lengthy negotiation, an agreement was struck and a draft contract, prepared by Nielsen on his laptop, signed. This was followed by a visit to Latcharter's office where another agreement regarding the delivery of the aircraft was negotiated. It was estimated that no more than two weeks would be required to bring the aircraft up to full state of airworthiness and that the work could commence as soon as the deposit was paid. Nielsen arranged payment of a 10 per cent deposit ($25,000) by direct transfer from Hong Kong to Latcharter's bank account in London. He worked this out using Masagutov's office telephone. Bleach and Nielsen agreed to return in two weeks to accept delivery of the aircraft. An express condition of the sale was that the air crew would be hired locally.

On Thursday, 2 November 1995, Bleach returned to the UK on a Latcharter jet. Nielsen had told him that he had business in Europe over the next two days. However, Bleach learnt that Haestrup had joined Nielsen at some stage, after he received a telephone call from the two while they were driving somewhere near Vienna. Nielsen was back in Hong Kong on 5 November 1995, but five days later he attended a meeting with the board of directors for a Hong Kong-based company, Carol Air Services Ltd.

Carol Air Services Ltd was discovered to be one of several shell companies that were floated before the actual arms drop, leading some senior Indian intelligence operatives to suggest that creating a maze of fictitious companies was an old intelligence trade craft that had previously helped some Western intelligence organizations, especially the CIA, to carry out a number of covert operations in several parts of the world at the height of the Cold War, especially in the 1970s and eighties. Partly because of lack of expertise in unravelling clandestine multinational intelligence operations and partly because it wanted to remain focused on the criminal case arising out of the arms drop, the CBI did not vigorously pursue these companies. Neither did it try to find out Nielsen's precise relationships with the executives of the companies.

It is interesting that Carol Air Services came into being slightly over two months before the 17 December 1995 Purulia arms drop. A copy of the minutes of a meeting held at the company's headquarters in Suites 1-3, Kinwick Centre, 32 Hollywood Road, Central, Hong Kong, on 11 October 1995, clearly says that the company was registered in the tax haven, Turks and Caicos Islands, on 6 October 1995.[13] All of the 5,000 shares of Carol Air Services Ltd were issued to a company named Owl Investments Ltd of Sovereign House, Station Road, St Johns, Isle of Man. Owl Investments Ltd also held all 5,000 shares (at $1 per share) of M/s Well Known Investment Corporation. Thus, Carol Air Services Ltd and Well Known Investment Corporation were fully owned subsidiaries of Owl Investments Ltd[14], which declared Wai Hong Mak as a nominee for the shares of Carol Air Services Ltd and Nielsen (or Davy) as the nominee for the shares of Well Known Investment Corporation. At the meeting on 10 November, Nielsen was given the power of attorney, which authorized him to execute acts on behalf of the company. Carol Air Services Ltd would ultimately own the AN-26 aircraft.[15]

Meanwhile, Bleach booked another ticket on a flight to Riga for 13 November 1995. Arriving in Riga, he checked into Hotel Latvia where he met Nielsen, who had arrived earlier. The next morning the two went to the offices of Latcharter where they met Viktor Masagutov. They found that no work at all had been done on the aircraft and it was not in any way ready for delivery. When the three visited the offices of Latavio Latvian Airlines, they found that they had to deal with a new management. In their absence, the company had been taken over by a firm of attorneys. The previous management had all but disappeared and the new management wanted to renegotiate the contract and put up a fresh price. This was bad news for Nielsen. 'A fairly acrimonious meeting followed, after which we left and returned to our hotel with Viktor in company,' Bleach wrote in his statement to the CBI.[16] A sheepish and embarrassed Masagutov assured Bleach and Nielsen that it was all a misunderstanding and, given a bit of time, he would be able to sort things out. Still fuming, Nielsen accepted Masagutov's assurance, so the three returned to Latcharter's office to make further plans. Following consultations with Latcharter's chief navigator and chief pilot on the delivery flight to Bangladesh, it was decided that Nielsen and Bleach would use the services of Baseops Europe Ltd[17] for the first flight from Riga to Dhaka on 22 November 1995. Baseops had the reputation of being efficient and competitively priced.

Bleach immediately called up Baseops Europe Ltd, where he spoke to one Sandy Love, and it was arranged that the company would carry out all flight arrangements for the aircraft. A fax message from Bleach to Baseops on 6 November 1995 is revealing. It laid out the following flight plan:

> Could you please make the necessary arrangements for a flight for the above mentioned aircraft (AN-26, "YLLDB") as follows:
>
> Fri 8th December KARACHI TO SHARJAH Overnight stay
>
> Sat 9th December SHARJAH TO BURGAS (Bulgaria) with technical stop at TURAIF (Saudi Arabia)
>
> Overnight 9th and 10th at BURGAS
>
> Mon 11th December BURGAS TO KARACHI, with technical stops at SIIRT (Turkey) and YAZD (Iran) Overnight at KARACHI
>
> Tue 12th December KARACHI TO RANGOON, with technical stops at VARANASI (India), DHAKA (Bangladesh)
>
> Thu 14th December RANGOON TO KARACHI, with technical stops at DHAKA and VARANASI.
>
> Purpose of visit to Rangoon is delivering passengers for business meetings.
>
> If possible, we would like this flight to have the number CAS-101
>
> The aircraft will be flying with its usual crew, including Mr Peter Davy, and I will be joining the aircraft in BURGAS and remaining with it until the journey is completed in Karachi where I will leave the flight and return to the UK via commercial (flight).[18]

Nielsen also spoke to Sandy Love and arranged a cash deposit to be paid against the costs. Once the payment of $5,000 was made, possible flight plans were drawn up. The Latcharter navigator strongly recommended that a Global Positioning System (GPS) be obtained so it could be fitted to the YL-LDB. Since there was little for Bleach to do, it was decided that he might as well return to the UK and purchase a GPS. Nielsen, meanwhile, would stay back to finalize the purchase and hire a crew. He handed out $600 (of $100 denomination) in cash to Bleach for purchase of the GPS.

Bleach landed at Gatwick Airport on 15 November 1995 and took the Gatwick Express train to Victoria Station before arriving at the Transair 'Pilots' Shop' adjacent to the station. Here he purchased a Trimble GPS and then proceeded homewards. Back in Riga, a purchase contract for the AN-26 was signed between Nielsen, representing the buyer, Carol Air Services Ltd, and Mara Bekere, representing the seller, Latavio Latvian Airlines. The cost of the aircraft was $180,000. According to the contract, the buyer would make a payment by bank transfer to an account at the Swiss Bank Corporation in Zurich, Switzerland. The beneficiary bank was Parekss Bank Corporation and the money was to be deposited in favour of Delaware-based JLY Enterprises Corporation (whose director was a US citizen, Eric Lee), which acted as the agent of Latavio. Long before the aircraft was purchased, the mysterious Eric Lee appeared on the scene and met the general director of the Latvian airlines, Aivars Rittenberg, claiming that he represented JLY Enterprises Corp. 'and offered to find a purchaser for an aircraft as a mediator'. Lee, who spoke 'English and broken Russian', 'explained' that he was aware of the 'financial difficulties' Latavio was in and had brought along in advance a 'copy of the agreement on cooperation' which was signed on 30 October 1995. On 2 November 1995, Eric Lee 'addressed' Rittenberg a 'preliminary prepared contract order, mentioning that [the] Antonov-26 aircraft, YL-LDB, should be sold at a price of $160,000, not lower, and with prepayment'. Rittenberg signed the contract 'not paying attention to the fact that it had already been signed' on behalf of JLY Enterprises Corp. by a lady, Mara Bekere (a law student), who was unknown to him. 'On November 15, 1995, Kim Davy, as he introduced himself, came to me with the contract on sale,' recalled Rittenberg.[19]

Four days later, Bleach received a phone call from Nielsen, who said that he had resolved all the problems surrounding the aircraft and that its sale would be completed the following morning, on 20

November. Nielsen told Bleach to meet him at Plovdiv Airport in Bulgaria, where the arms would be loaded onto the AN-26 and the aircraft refuelled. He informed Bleach that Baseops had booked the crew at the Plovdiv Novotel Hotel and that the next meeting would be there. He left Riga on the morning of 21 November.

Like a dutiful partner-in-crime, Bleach also took a flight from London to Sofia on 21 November and then hired a car to drive to Plovdiv, a distance which he covered in less than two hours. When Bleach checked into Plovdiv Novotel, there was no sign of Nielsen, who finally reached around midnight. With him were seven Latvian crew members. The delay, apparently, was caused by a last-minute hitch in the purchase of the aircraft. Although the money had been transferred from Hong Kong to Latvia, the new management of Latavio wanted to make sure that the money had duly reached the account before parting with the aircraft. Nielsen had had to arrange for a coded telex to be sent to Hong Kong to fix the purchase glitch.

That very day, without wasting much time, Nielsen flew out of Riga, with Dhaka as the destination, on the aircraft commanded by the pilot-in-command, Alexander Klichine aka Sasha. The other members of the crew were: Oleg Gaidach (co-pilot), Igor Timmerman (flight engineer), Igor Moskvitin aka Alexandre (navigator) and Evgueni Antimenko (operator). There were two ground crew members, Vladmir Ivanov and Alexander Lukin. Nielsen signed an employment contract with the crew for a period of three months. The contract stipulated that an amount of $2,217 per week would be paid to each, besides other benefits like accommodation, transport, medical facilities, etc. Bleach could not meet the crew that night because they had exceeded their flight time with the delay and had gone directly to bed. He would have the opportunity to meet the crew later. That day—21 November 1995—the AN-26 was struck out of the aircraft register by the Civil Aviation Authority of Latvia[20] and a special permission for the flight 'Riga–Dhaka' was issued to transport the aircraft to a new base. But it was not re-registered in any other country.[21]

**Alexander Klichine, aka Sasha,
the pilot-in-command.**

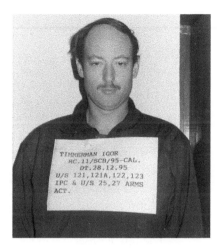

Igor Timmerman, the flight engineer.

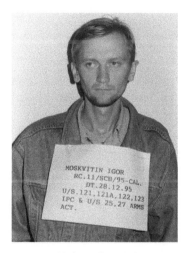

Igor Moskvitin, the navigator.

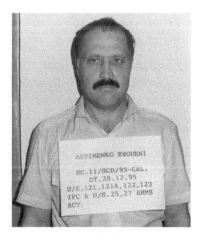

Evgueni Antimenko, the operator.

Oleg Gaidach, the co-pilot.

The Arms Are Procured

Because of the nature of the arms needed and the constantly changing requirements, very little was done to obtain the necessary equipment until very late in this affair. Bleach's first step to actually assist Nielsen to procure the weapons did not take place until the second half of December, when he called up a number of suppliers to check on the availability at short notice. He found that procurement would be no problem at all.

The initial order was for 2,500 AK-47 Type 56 Mk-2 rifles and 1,500,000 rounds of ammunition. After the Bangkok meeting, the order was reduced to 500 of the same type of assault rifles and 250,000 rounds of ammunition. Later still, sometime during late November, the order was changed again, this time the consignment that was actually delivered, minus the pistols: 300 AK-47 rifles, 250,000 rounds of ammunition, 100 rocket-propelled anti-tank grenades, 6,000 bullets for 9-mm Makarov pistols, 100 offensive grenades, 25 PM-79 anti-personnel landmines, 10 RPG-7 rocket launchers, Dragunov rifles with optical sights and some night vision equipment. Bleach was surprised at how the pistols came into the picture, but did not worry much about it at the time.

Bleach claims that he attempted to dissociate himself from the actual arms cargo around that time.[22] He contacted one Samuel Sieve, a London-based broker of arms deals heading a company called Trade Investment UK Ltd, and a few others, to put Nielsen directly in touch with them. As Nielsen's flight plan called for a refuelling stop somewhere near Bulgaria, that would be an obvious place to deliver the arms before they were loaded onto the aircraft. Arrangements were finally made and it was confirmed that the required weapons would be supplied and ready-packed for transport at an airport in Bulgaria. Nielsen was told that he would be advised at which airport the arms would be made available, so that he could file his flight plan accordingly.

On or about Friday, 10 November 1995, Nielsen was advised that the consignment would be ready at Plovdiv airport and could be collected during the week commencing Monday, 20 November 1995. However, by Friday, 17 November, it was clear that it would not, in fact, be ready by the date promised. Nielsen, naturally, was anxious to get the aircraft out of Riga before anything else went awry with the purchase and decided to keep to the original plan. On the morning of Tuesday, 21 November, Bleach received a telephone call on his mobile phone at Heathrow, just prior to boarding the flight to Sofia, with the caller saying that the consignment was definitely not ready and asking him to pass the message on to Nielsen.

When Bleach met Nielsen at Plovdiv Novotel Hotel that evening, he broke the news to the Dane. At first Nielsen was disappointed and a little angry. However, it then turned out that he had also purchased a large quantity of aircraft spares, all of which were on the aeroplane, and that he could not have carried the weapons anyway. As a result, Nielsen and the crew decided to fly to Dhaka and return for the arms at a later date. Suddenly, information came through that the arms were ready for shipment the following day, i.e. Friday, 24 November 1995. That day,

> the goods were checked by the gentlemen Mr (Peter) Scott and Mr (Robin) Campbell (of BTI Ltd) and Mr Bleach in the presence of the representative of KAS Engineering Co at the manufacturing plants of Arsenal and VMZ... A certificate of inspection regarding the acceptance of the goods was signed by Mr Robin Campbell[23].

Exhausted, Bleach returned to his North Yorkshire home, and spent the following week in bed after being taken ill with a severe attack of flu. On Monday, 4 December 1995, he received a call from Nielsen informing him that Nielsen was ready to return to Europe and collect the arms. He said that the delivery was to be at Burgas Airport, on the shores of the Black Sea, in Bulgaria. He told Bleach that he expected to arrive on Sunday, 10 December, and gave him a provisional flight plan for YL-LDB. He then asked Bleach to arrange for refuelling, etc. with Baseops, which the Briton carried out immediately.

On 10 December 1995, Bleach flew British Airways from London's Gatwick airport to Sofia, where he took an internal Balkan flight to Varna, from where he drove to Burgas, arriving shortly after midnight.

He checked into Hotel Cosmos, which gave discounts on tariff to crew members. Nielsen and the crew had already arrived. When Nielsen told Bleach that the arms had been loaded onto the AN-26 aircraft that morning, the Briton was surprised and wondered whether the crew members, all ethnic Russians who lived in Latvia, had witnessed the loading.

Providing details on how the arms reached Burgas, Nielsen told Bleach that he had initially paid a deposit of $50,000 by direct bank transfer to a Swiss bank account and that the balance was paid in cash by bankers' draft at Burgas Airport on the morning of 10 December 1995. There was no formality whatsoever at Burgas and the arms and ammunition were simply delivered on a truck and loaded straight into the aircraft's rear cargo hold. The weapons, packed in wooden crates, were consigned as 'technical equipment' and Nielsen was in possession of the consignment note to that effect. The boxes were stacked in the rear of the aircraft, covered with blankets and secured with cargo straps. Nielsen showed Bleach a packing list which gave the dimensions and weights of all the boxes (but not the contents). The total weight came to 4,375 kilograms. The total cost was $165,000 and, therefore, the balance of the bankers' draft which he had handed over was in the region of $115,000.[24]

Weapons' Source and the End-User Certificate

Bleach chose not to mention anything on how the arms were actually procured in his statement to the CBI. However, some documents— essentially English translations of the interrogation reports of some Bulgarian officials—revealed the process and the companies that were involved before the arms and ammunition reached Burgas Airport. Russy Houbanov Russev, executive director of KAS Engineering Co., a conglomeration of trading firms belonging to the defence industry complex of the Republic of Bulgaria, said during interrogation that the 'deal realized' was for the export of 'special production' (of weapons) for Bangladesh's Ministry of Defence through a London-based company called Border Technology and Innovations Ltd (BTI). The weapons and the ammunition were manufactured by two Bulgarian firms, Arsenal Ltd[25], based in the town of Kazanlak, and Vazov Machine-Tool Producing Plant, in the town of Sopot. He revealed that the deal was realized on the basis of a correspondence between BTI, two of whose representatives, Peter Scott and Robin Campbell, spearheaded the negotiations. Russev continued:

> The correspondence on the present deal started with a letter dated November 8, 1995, which was in fact a request for the following items: RPG-7 (rocket-propelled grenade), RM-79 (landmines), 7.62 AKS-47M, 9mm Makarov revolvers, Dragunov sniper (rifles), night vision equipment, grenades, 7.62 x 39 ammunition, 9 x 18 mm bullets and anti-tank rockets PG-7... Before signing the contract we received the original copy of the end-user certificate No. 4021/1/AA/ARMY/ASL(P)/2 dated November 9, 1995, duly signed and sealed by the Chancery of the Prime Minister of the Republic of Bangladesh, Military Department, and also an authorization letter dated November 25, 1995. The end-user certificate shows that the items mentioned are for the army of Bangladesh, that they are not subject to export to other countries and that BTI is the company authorized to do the deal.[26]

The EUC, which Syed Tasleem Hussain is suspected to have procured, appears to have been typed out on plain paper and not on any official stationery of the Bangladesh government, giving rise to the suspicion that it was a forgery. The only sign of any official stamp is the issuing signature of a senior officer, Major General Mohammad Subed Ali Bhuiyan, principal staff officer, Prime Minister's Office, Armed Forces Division, Dhaka Cantonment. This is how the EUC, issued on 9 November 1995, begins: 'We hereby confirm that the following products purchased under Contract No. 214/719/Project/DGDP/Ord/P-4 dated 14 September 1995 will be used by the Bangladesh Army and will not be exported to any other country.' It was suspected that Major General Bhuiyan accepted a generous bribe to sign on the EUC, a matter that was later investigated by the Bangladesh government when Sheikh Hasina became the prime minister, following the 1996 general elections in which the Bangladesh Nationalist Party (BNP), which the army officer was close to, came to power.[27] Major General Bhuiyan was cashiered and forced to take early retirement. In all likelihood, he received a tidy pay-off from Syed Tasleem Hussain to sign on the EUC. Once the relatively 'friendly' Awami League government of Sheikh Hasina Wajed was in power in Dhaka after the 1996 general elections, the R&AW was able to compare the EUC used for procuring the Purulia consignment of weapons with other EUCs issued by the Bangladesh Armed Forces Division. It found that BTI's name was not even mentioned in the actual EUC as the supplier of the weapons. The R&AW also discovered that Syed Tasleem Hussain's 'success in business was attributed to patronage of the then Bangladesh Prime Minister Begum Khaleda Zia' whose brother Syed Iskander had a stake in Hussain's company, Riverland Agencies Ltd.[28]

The weapons listed in the EUC included 300 rifles, automatic 7.62 mm x 39 Type AK-47, folding metal stock; two rifles, semi-automatic, 7.62 mm x 54 Type SVD c/w optical sights; twenty-five pistols, 9 mm x 18, 'Makarov', Type PM; ten rocket launchers, Type RPG-7; twenty-five mines, A/personnel, Type PM-79; 100 grenades (offensive); 100 grenades, anti-tank, Type PG-7V; 25,000 7.62-mm ball ammunition; 6,000 9-mm ball ammunition, two binoculars, night vision; and two sights, rifle, night vision. A subsequent secret letter (25 November 1995) of the Bangladesh Directorate General Defence Purchase (DGDP), signed for the director general by a Major M. Jasim Uddin, however, authorized M/s Border Technology and Innovations Ltd to 'take all steps to conclude the subject contract, namely 214/719/19/Project/DGDP/Ord/P-4'[29]. The wooden boxes containing the arms and ammunition bore the seal (in stencilled capital letters) 'COMMANDANT CAD RAJENDRAPUR CANTT, BANGLADESH'. On one of the wooden boxes was written in bold: 'CASE NO. 34 OF 60, CONTRACT NO. 214/719/ PROJECT DGDP'. Some of the markings were in the Cyrillic alphabet.

Several months after the arms drop, BTI's Peter Scott declared in his witness statement to the British police on 2 August 1996 that his company had received the EUC from Bleach. A message dated 13 November 1995, signed by Scott, was sent by DHL courier service, addressed to Russy Houbanov Russev. An EUC was apparently enclosed which 'BTI received from the representative of the end-user'[30]. Scott asked Russy Houbanov to submit the EUC to the Bulgarian authorities for validation and subsequent export licence approval. Was Bleach the representative of the end-user? From all available indications, it does become apparent that Bleach used his contacts in the Bangladesh military to procure the EUC and was, perhaps, doing a service to Nielsen who, it appears, did not have any sources in the Bangladesh military establishment to have procured the EUC. Whether the EUC was forged or not, it did convince KAS Engineering Co. and other Bulgarian authorities, who gave the clearance for the loading of the wooden crates containing the weapons and the ammunition onto the AN-26 aircraft (on 10 December 1995) at Burgas Airport.

Four days before the arms and ammunition were loaded onto the AN-26, Peter Scott sent a fax to KAS Engineering Co., informing Russy Houbanov Russev that the shipment from Burgas had been arranged and that he had attached the flight and packing details. Scott added that he would fly out to Sofia on 9 December and sought a rendezvous for 10 December 1995 at the Moscow Hotel, Sofia. Scott suggested, 'We can meet at the hotel at 1000 hours to await the result of pre-shipment checks and then to give authority for (the) release of good once we have confirmation from London that payment has been received by BTI.'[31]

The very day Scott sought a rendezvous with Houbanov, BTI's Robin Peter Campbell met a person in London described as being an Irishman called 'Mick'. The meeting took place at St Giles Hotel, Bedford Street, London. Mick had brought an envelope containing two bankers' drafts drawn on HSBC Bank, totalling $70,000. He was not allowed to release the drafts without authorization. Later that day, Campbell was asked to meet Mick in his hotel room to collect the drafts. Shortly after entering the room, Nielsen telephoned from Bulgaria and authorized the release of the payment.[32] Mick would later be identified as Leonard Paul Francis, alias Perry Mick Joseph, born on 9 December 1956 in Dublin and Nielsen's former partner-in-crime. He was sentenced on 25 October 1983 to serve a four-year imprisonment and was deported by the Gothenburg district court in Sweden to the UK for a heist at a jewellery store in April the same year.

6

A STRUGGLE, BUT CARGO AWAY

While Peter Bleach lay convalescing at his Howdale Farm home in North Yorkshire, in the third week of November 1995, the AN-26, with the call sign YL-LDB, flew out of Plovdiv Airport in Bulgaria on 22 November, headed for Dhaka. On board were Alexander Klichine (the pilot), Oleg Gaidach (the co-pilot), Igor Timmerman (the flight engineer), Igor Moskvitin (the navigator), Evgueni Antimenko (operator) and Kim Palgrave Davy, alias Niels Christian Nielsen. There were also two Russian ground engineers.

The maiden flight of the AN-26 under its new owner did not have a seamless start. At the time of departure at 1000 hours, Bulgaria time, on 22 November 1995, permission for landing had not been confirmed by the Bangladesh Controller of Certifying Authority (CCA). It was not unusual to take off before the final leg of the flight was confirmed, as usually this is a formality. The route followed would take the aircraft via Iran, with refuelling at Tabriz and another location, and then on to Karachi in Pakistan, where there would be a compulsory overnight stop. From Karachi, the AN-26 was to fly on to Dhaka the next day and refuel at Varanasi, in Uttar Pradesh. The flight encountered no problems and touched down safely at Karachi airport on the night of Wednesday, 22 November. There, Nielsen was met by the enthusiastic staff of Shaheen International, the ground handling agents, who had earlier been arranged by Baseops. Shaheen had booked the crew into Pearl Continental Hotel, which gave discounts to aircrew.

The next morning, Nielsen and the Latvians found that Dhaka

had still not issued landing permission. They contacted associates of the proposed joint venture company, including, perhaps, Syed Tasleem Hussain, who assured Nielsen that the permission had all but been arranged. Nielsen, therefore, decided to take off as planned and check again with Dhaka once they landed at Varanasi. On reaching Varanasi, they found that landing permission at Dhaka's Zia International Airport had still not been granted. Nielsen again called Dhaka and was assured once more that the landing permission was in hand. So it was decided to spend some time in Varanasi and await the permission. They checked into Hotel India in Varanasi, where the seven men (Nielsen, the five Latvians and the two Russian ground engineers) ended up staying on for at least four days, during which time Nielsen sent out the Latvians for a number of sightseeing trips, including a boat ride on the Ganges. Nielsen also took the opportunity to meet a few people in Varanasi, besides closely observing the security arrangements as well as the alertness level of the airport staff. The flight to Varanasi and the sojourn at the holy city was, after all, a reconnaissance mission. Between 24 and 26 November, Nielsen effected three money transactions on his Visa account for payment to hotels in Varanasi.[1] While reconnoitring Varanasi, Nielsen contacted two Russians, Mikhail Kalyukin (passport no. 43N-1977217), the Khabarovsk-based man who had sent across two flight engineers to Riga to inspect the AN-26 in the middle of November 1995, and Dovard Chikanov (passport no. 43N-6166279), who had put up in Room No. 608 of Bay View Hotel in Singapore.[2]

The Dhaka leg of the flight could not be completed as the landing permission never came through. Intelligence gathered by the R&AW suggested that the Bangladesh Directorate General of Forces Intelligence (DGFI) had either suspected or had specific information on the nature of the flight and the cargo it intended to carry.[3] In fact, Nielsen's Carol Air Services Ltd had approached the Civil Aviation Authority of Bangladesh on 19 November 1995 for landing permission, claiming that the aircraft would carry general cargo from Shaheen Airport Services for Air Parabat Ltd, Bangladesh. Carol Air Services provided its UK postal address as that of Baseops Europe. Responding to a query from the Bangladesh civil aviation authorities, Carol Air

Services claimed that it sought landing permission so it could showcase a static demonstration of the AN-26 at Dhaka's Zia International Airport. It named Bangladesh's just-retired chief of air staff, Air Vice-Marshal Altaf Hussein Choudhury, as a local reference. In the event, a letter from the then Bangladesh Civil Aviation assistant director, Mohammad Habibullah, to Dhaka-based Air Parabat Ltd, associated with Syed Tasleem Hussein, said that 'no action can be taken' on the landing permission request.[4] Suspecting something fishy, the DGFI recommended that no landing permission be given to the AN-26. So the aircraft returned to Karachi on 28 November. Bleach suspected, not without reason, that over the next few days, the aircraft flew to Sharjah. Nielsen had telephoned him from a Sharjah number on one of those days and given him a proposed route to Burgas to file with Baseops, which Bleach did before duly meeting Nielsen and the crew at the Bulgarian airport on 10 December. These flights were later found to be 'dry runs', to test not just the aircraft but also the alertness levels of the Indian authorities.

The Second Flight

When Bleach arrived at Hotel Cosmos in Burgas around midnight on Sunday, 10 December 1995, he found that the crew had gone to bed. He was met at the hotel lobby by Nielsen who disclosed that the arms and ammunition had been loaded onto the AN-26 that morning.[5] Altogether seventy-seven cases, each marked 'technical equipment', were loaded. 'The entire operation was smooth,' Bleach recollects Nielsen telling him. The arms had all been boxed, stacked evenly in the freight compartment of the aircraft, covered with blankets and secured with freight fittings. Nielsen said there were no formalities at all at the airport and that the goods were loaded quite openly in full view of the people around. The plan was to leave early the next morning, so Nielsen and Bleach retired to their rooms immediately.

On 11 December 1995, Nielsen[6], Bleach and the five crew members reached Burgas Airport around 0600 hours, but they encountered some difficulty before gaining access to the aircraft. A flight plan was filed for Dhaka as the end destination, with two refuelling stops in Iran and an overnight stop at Karachi. For the next leg, it was decided that refuelling would again be at Varanasi before landing at Dhaka. Nielsen said that he wanted to fly to Phuket to give the crew a Christmas break. The arms would be delivered between Varanasi and Dhaka. Like its maiden flight from Plovdiv, this time, too, the AN-26 encountered a few niggling problems.

At Burgas Airport, Nielsen, Klichine and Moskvitin went to the air traffic control while the rest walked to the aircraft. However, that day, the Bulgarian CCA decided to check all documents and it was found that the Certificate of Competence for the AN-26 was missing. The flight would not be allowed to take off without the document. The plane returned to Sofia. Bleach 'departed' Bulgaria as he had entered the day before, and his passport had an entry stamp. However, on 're-entering' Bulgaria, he was placed on the crew document and so did not receive an individual stamp, as is normal for aircrew. From this point on, Bleach's passport did not reflect details of his travels.

After checking back into Hotel Cosmos, Nielsen telephoned Latcharter's Victor Masagutov in Riga, Latvia, and told him about the problem over the Certificate of Competence. It turned out that this document should have been supplied with the aircraft but had been retained in Riga by mistake. Masagutov undertook to contact Latavio and try to obtain the document. Burgas air traffic control had said it would be happy to accept a faxed copy. Later that day, a copy of the document arrived by fax at the Burgas air traffic control, which notified Nielsen that permission to take off the next morning (12 December 1995) had been granted. There was a round of cheering and collective back-patting.

Around 0730 hours the following morning (Tuesday, 12 December 1995) Nielsen, Bleach and the crew returned to Burgas Airport and moved to the airside[7], again with no passport stamp. As usual, Nielsen, Klichine and Moskvitin went to the air traffic control to check the flight plan and weather conditions, while the rest walked to the aircraft. There was some delay with the flight plan and it was not before a couple of hours that the AN-26 was able to take off. YL-LDB was airborne with the lethal cargo in its dark, cavernous belly.

The distance of 1,038 miles from Burgas to Tabriz was covered in two hours and twenty minutes. According to its flight plan, the aircraft was scheduled to make a technical stop for refuelling at Gaziantep in Turkey, but flew direct to Tabriz because of the shortage of fuel at Gaziantep airport.[8] At Tabriz, the cargo for the flight was declared to be '9,000 kilos of non-commercial aircraft spares'[9]. After the AN-26 was refuelled, the crew was informed that there was a serious weather problem at the proposed next stop so the aircraft would be diverted to Ispahan, which was to be strictly a technical stopover for refuelling. However, when the aircraft landed at Ispahan, a thick layer of fog covered the airport, bringing visibility down to only five metres. Under the circumstances, air traffic control denied the crew permission to take off. The aircraft, with seven persons on board, remained on the tarmac for the entire night. No security guard checked the interior of the plane, nor did anyone ask about the cargo it was carrying. As usual, an armed guard was placed outside the aircraft, but no more than that.

When the fog lifted next morning (Wednesday, 13 December 1995) around 0800 hours, the AN-26 was given permission to start the engines and take off for Karachi. However, with the temperature well below zero degrees, the aircraft battery had drained overnight because of the use of power to talk to the control tower. The crew was unable to start the engines and Klichine had to call for ground power, which was brought over from the nearby air force base. As a result, it was past 1100 hours when the aircraft finally took off for Karachi. Ispahan is a very difficult airport, completely surrounded by mountains and with a complex approach pattern. Both landing and take-off were turbulent.

The 1,072-mile distance between Ispahan and Karachi was covered in two and a half hours, with the AN-26 aircraft landing at Qaid-e-Azam International Airport in the late afternoon. It was met by the Shaheen International ground staff, parked and a policeman was placed on guard. Shaheen personnel took Nielsen, Bleach and the crew through Customs and immigration and then drove them to Pearl Continental Hotel. The pilots had now been in the cockpit in excess of twenty-four hours—way beyond the maximum permissible time—and were badly in need of sleep. After an early dinner, everyone retired to bed.

On Thursday, 14 December 1995, everybody spent the day by the hotel swimming pool, doing nothing in particular. No business was discussed and everyone, including Nielsen, simply relaxed. A courier flew in from Hong Kong to deliver $50,000 in cash to Nielsen. In the evening, a hospitable Shaheen staff introduced the Latvians to some Russian girls living in Karachi. They took the girls out shopping and bought them expensive gifts. For Nielsen and Bleach, the big day was nearing.

The Latvian crew members and Nielsen relaxing by the pool at Pearl Continental Hotel, Karachi.

The next day, Nielsen and Bleach prepared the temporary crew identification tags, which were subsequently replaced by new, more authentic-looking ones that had been made in Hong Kong and sent to Nielsen by courier service. Nielsen was unable to find a colour ink-jet printer or a colour laser printer and so had to make do with a colour dot matrix printer, which was why the quality of the cards turned out to be so poor. The lamination was done at the hotel's business centre. Later that afternoon, Nielsen held a meeting in his room with one Farooqi, the general manager of Shaheen International, with a view to discussing a proposed joint venture freight company. (Nielsen had decided he would not do business with the Bangladeshis. He had become keener to enter into a partnership with Shaheen.) Around 1600 hours, Nielsen called Bleach at the swimming pool and asked him to join in the meeting. Bleach was introduced to Farooqi, who wished to know whether he would be able to find a suitable freight agent at Heathrow Airport. Bleach said he did not think that would be very difficult and there followed a general discussion on the prospects of air freight in the region. Farooqi left around 1730 hours, and Nielsen and Bleach continued with the discussions. At that moment, Nielsen pitched the proposal whether Bleach would be interested in taking part in the joint venture freight company.

But Bleach had other important things on his mind. 'The time had come to settle our business so far,' Bleach wrote in his statement to the CBI.[10] Nielsen had offered to pay Bleach 10 per cent of the aircraft purchase price as commission for his work and Bleach proposed that if he offered him a spares contract for two to three years, he could consider reducing that percentage accordingly. He told Nielsen that as he had now delivered the aircraft and the flight trials had gone satisfactorily, the time had come to settle up. Nielsen said he was running slightly short on cash and that he had the hotel bills and fuel to pay for. He said that since he would receive fresh supplies of money in Phuket, it would be best if Bleach accompanied him to the Thai resort town, where the Briton could enjoy a free holiday at the Dane's expense. An extremely good hotel had been booked and all expenses would be on Nielsen. Bleach could fly back to Karachi with Nielsen, collect his commission and return home on a business-class air ticket well in time for Christmas. Little did Bleach suspect that the wily Dane had already ensnared him and ensured that the Briton did not abandon the project when it was at its most delicate stage.[11]

Bound by Nielsen's proposition, Bleach returned to the subject of the cargo, pointing out that since there were no parachutes, it seemed pointless flying out again as the AN-26 would not get permission to land at Dhaka. (Peter Haestrup's attempts to sell a cover story to the Bangladesh Civil Aviation Authority that the AN-26 would be used to transport tiger prawns from Bangladesh to Hong Kong had come to nought. When he tried offering a bribe, suspected to be $50,000, a senior Bangladesh Air Force officer baulked at the huge amount being offered as grease money, suspecting something extremely fishy.) Bleach suggested holidaying in Karachi and setting up the freight company there at the same time. He told Nielsen that it would be quite feasible to repaint the boxes containing the arms and ammunition and leave them in a warehouse with the aircraft spares so that they could be dealt with at a later time. It was then that Nielsen let out a small secret, that a friend would arrive from South Africa shortly, bringing with him the parachutes.

The Parachutes Reach Karachi

On Saturday, 16 November 1995, Bleach accompanied Nielsen into Karachi and went to the timber and wood-working area of the city. Nielsen had brought along with him the dimensions of the aircraft's cargo rails and ordered a total of three pallets to be made. He left Bleach at the carpentry shop to supervise the making of the pallets. When they were done, Bleach paid the carpenter $300 before taking a taxi back to the Pearl Continental.

Deepak Manikan, alias Daya M. Ananda.

When Bleach returned to the hotel, Nielsen was not around but had left a message for him, saying that he was away at the airport to receive a friend. Around 1930 hours, Bleach accompanied the crew to dinner, and around 2100 hours Nielsen arrived with a man who was introduced as Deepak[12]. Deepak joined the group for dinner, where it was obvious to Bleach that like Nielsen and Randy, the newly arrived associate was a strict vegetarian.[13]

No business was discussed over food, but after dinner the three— Nielsen, Bleach and Deepak—went to Bleach's room to discuss the parachutes. In the midst of this conversation, Deepak said he had flown in from Johannesburg, where he had purchased the parachutes from a South African army surplus dealer. They were apparently used but were guaranteed for twenty more drops and ready-packed for use. The parachutes had been supplied complete with cargo nets and collapsible shock-absorbing bases. Deepak quickly drew a diagram to show how the cargo nets worked and how they were connected to the parachutes. Each of the parachutes had cost $5,000, and as each weighed about 75 kilograms in their boxes, Deepak had had to bring them to Karachi via Bahrain as excess baggage. Even more interestingly, the parachutes were not brought through Customs into Pakistan, but were taken straight to the aircraft by the ground staff of Shaheen International, who were told that these were 'industrial samples' bound for Bangladesh. [14] Bleach believed that since YL-LDB was in transit, it was normal for freight of all types to be transferred from plane to plane airside without Customs formality.

From the conversation, Bleach understood that Deepak knew perfectly well what the parachutes were for and was fully conversant with the entire operation. It became clear to him through the evening that Nielsen and Deepak had known one another for a very long time and that Deepak was thoroughly familiar with and deeply involved in the gold smuggling operation.

Nielsen and Deepak went to their separate rooms after arranging to meet at breakfast the following morning, 17 December 1995.

Next morning, around 0900 hours, Nielsen, Bleach, Deepak and the five crew members checked out of the hotel. Nielsen paid for all the bills with his Bank of America credit card (No. 4910-1715-4336-7601). The pallets were delivered to the aircraft by a pick-up truck supplied by Shaheen International, which was quite normal, according to Bleach, because most ground handling staff at airports across the world have prior security clearance and are usually not stopped or questioned during such routine tasks. Nielsen had provided the name and location of the carpenter to Shaheen and they had simply sent a vehicle to collect them. The pallets were simple wooden structures, marked YL-LDB #001, YL-LDB #002 and YL-LDB #003. They passed through Customs on entering the airport and were cleared in the normal course. There was no reason for the Pakistan Customs official to be suspicious. The eight men were ready to set out on their mission.

The pallets were loaded by hand through the rear cargo ramp and stacked leaning against the sides of the aircraft. They were secured with cargo straps. That the aircraft had freight on board was visible during this operation. However, the nature of the freight was not clear. Once again, Bleach claimed in his statement to the CBI, there was no reason for Pakistani officials to question this.[15] The aircraft was in transit and technically, being airside, was not even in Pakistan. The contents of the aircraft would only become the business of Pakistani officials if an attempt was made to unload the aircraft and bring the goods into the country. According to Bleach, 'this point has an interesting corollary in that during the period of [his] detention at Mumbai airport, it was pointed out to me very clearly by one of the interrogators that technically he was not even in India!'[16]

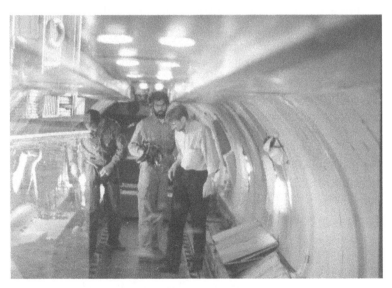

Deepak Manikan and Nielsen inspecting the aircraft before the weapons are loaded at Karachi airport.

At this time, Nielsen and Moskvitin attended to the flight plan at the Karachi air traffic control. As expected, landing permission had not come through from Dhaka and, therefore, it was decided to simply fly to Phuket as planned earlier. However, it was taken into account that flying from Varanasi to Phuket would not be possible without refuelling and so a request was made for a technical stop at Yangon, the Myanmar capital. At the time of departure during early afternoon, the request to Yangon had been acknowledged, though not granted, by the Myanmarese authorities.

Goods Are Fastened to the Parachutes

Immediately after the AN-26 took off from Karachi, Nielsen and Deepak began to work at a hectic pace to prepare the cargo for the airdrop. They were methodical and knew how to go about the business. The cargo occupied two-thirds of the available space, leaving a third of the space free. One of the new pallets was first laid on the rollers between the guides at the rear of the aircraft. It was secured with cargo stays. A parachute net was then laid flat on top of the pallet, and loading commenced. Deepak was in charge of this job, with Nielsen and Bleach giving him a helping hand.

With deft hands, Deepak removed the parachutes from their packaging. It was, of course, immediately obvious what they were, which caused some consternation among the crew. According to Bleach, the pilot, Alexander Klichine, 'was seething with rage but cowed by the weapons' that Nielsen and Deepak pulled out, threatening the crew to follow orders.[17] They were extremely unhappy and cursed away among themselves in Russian. It needed the calm Dane to pacify the Latvians by speaking to them in Russian, after which they cooperated grudgingly.

Because of the lack of space, work proceeded very slowly. While the cargo nets were rated at up to a ton, they were relatively small in dimension. Deepak decided that some of the boxes would have to be emptied and the contents placed loose on the nets. This would help with the weight because the boxes were extremely heavy. Bleach, therefore, helped Nielsen open some of the boxes while Deepak continued to pack the nets. Unloading the boxes actually made the problem worse because there was no place to put the empty boxes without taking up more valuable extra space. As the first pallet was filled, space was created in front of it, and it could slide forward on the rollers and allow the next pallet to be laid on the track.

The entire process was time-consuming and, as a result, by the time the aircraft began to approach Varanasi, the job was nowhere near complete. Once the plane landed at Varanasi, Nielsen asked Deepak, 'How much more time will you take to finish the entire job?'

Deepak replied, 'Well, I will probably manage in three hours.'[18]

Deepak did not want to do any work on the pallets between Varanasi and the drop zone because he wanted to use the time to rig the parachutes. He wanted to take the parachutes out of their packs and check that they were folded correctly. Meanwhile, Nielsen instructed Alexander Klichine to inform the air traffic control tower that the aircraft would be delayed for a short time for technical reasons. With the message conveyed, Nielsen, Klichine and Moskvitin left the aircraft to visit the air traffic control tower to report on the flight plan. A new flight plan, indicating Calcutta as the next destination of the aircraft, was filed with the air traffic control at Varanasi airport.

Once this job was taken care of, Nielsen telephoned Randy on his satellite phone, informing him that the aircraft would arrive, but would be about three hours behind schedule. Randy was told to ensure that everybody was kept inside the house (at the drop zone) to avoid danger from free-falling pallets. Klichine was asked to check on Yangon, which had not yet granted landing permission.

Meanwhile, on the aircraft, Deepak pressed on, feverishly filling the pallets. He did a neat job, making a hollow cube formation with smaller boxes and then stacking the AK-47 rifles within the cube for greater protection. Pouches and soft items from the boxes were used to further cushion the rifles. As each pallet was filled (with just over 1,000 kilograms of goods in each), the cargo nets were folded up and secured tightly. When complete, the package was cubic in shape and tightly bound together. All of the six persons remained on the ground for about three hours, by which time the job was completed. It was not possible to fit all the goods into the nets and, therefore, Deepak decided that the remaining boxes would simply have to be free-dropped with the pallets to get rid of them. At the rear of the aircraft now were three pallets that lay edge-to-edge on the cargo rails, sitting on the rollers, each secured with freight straps. On top of each was a cargo net, secured to the pallet only with very light cord, designed to break away as soon as the load left the aircraft. Along the sides of the aircraft were stacked empty boxes, and the full boxes which were to be free-dropped were stacked on the edges of the pallets.

The plan was that the pallets themselves would simply be a means to eject the cargo from the aircraft quickly. They were not secured to the cargo nets and would, therefore, fall away as soon the packages left the aircraft. This would reduce the load on the parachutes. The excess full boxes would also fall away, leaving the parachutes to drift to the ground unhindered.

The moment the plane took off from Varanasi around 2200 hours (during investigation later, an analysis by the IAF suggested that the AN-26 took off from Varanasi around 2100 hours[19]), Deepak checked the parachutes' packing one by one and then rigged them to each cargo net. The parachutes were opened by small drag chutes, which in turn were operated by a static line. As YL-LDB was a civil version of the AN-26, there was no provision for parachute dropping. So the static lines were fastened to the cargo hook on the hoist to the rear of the aircraft. During the preparation stage, the crew remained mostly in the cockpit and out of the way. When Igor Timmerman and Oleg Gaidach were asked by Nielsen to help move the boxes, they did not protest.

The Drop

The drop zone was now very close to the air route from Varanasi to Calcutta, so the necessary deviation was minuscule. As soon as the AN-26 deviated from its flight path (R-460) over Gaya, it lost height and the final preparations were made. Everything moveable was stacked at the front of the aircraft, in the passage by the main entrance door and around the toilet. It was all secured with ropes and cargo nets. As the aircraft levelled out at its new altitude, most of the stays securing the pallets were removed, leaving just a couple for safety. Deepak produced mountain-climbing harnesses from his case, and both Nielsen and he put them on. They attached short safety ropes to the harnesses and attached the ropes to cargo securing points at the rear of the aircraft. Nielsen stationed himself on the right hand side (as you look towards the rear of the aircraft) and Deepak positioned himself on the left. Nielsen stood by the controls to the rear cargo door. Deepak wore the headphones which allowed him to listen to the cockpit radio transmission and messages from the air traffic control. Nielsen, Deepak and Bleach did not exchange too many words. They were tense, but in complete control. If Nielsen had any thoughts about his thirty-fourth birthday, which was minutes away, he kept them to himself.

Nielsen took out the handheld GPS unit to calculate the exact point of the drop, which was probably when a mistake was made at the very last moment. Since the grid reference (86.20 degrees east and 24.30 degrees north) had been programmed into the GPS, the GPS would simply count to zero as it approached the actual grid reference point. The aircraft's diversion would take it directly over the site, or so it was hoped. However, unknown to Nielsen, the pilot, whom he had instructed to cross the drop zone at 1,500 feet, had set the height at 1,500 metres.

While Deepak and Nielsen made the final preparations, Bleach finished securing all luggage and moveable objects. Once done, he tied a ten-feet-long rope around his waist, which he secured with one of the metal stays on the shelving next to the toilet. Nielsen then began a countdown with hand signals. As the cargo door opened, Bleach braced himself at the door frame of the toilet. The aircraft took on a distinct 'nose-up' tilt and at the very instant, the pallets were dropped. The first two went out quickly. Bleach could not see them but as each went out, the aircraft shook. The third pallet appeared to be stuck. It took twenty seconds for it to come free, but by this time the aircraft was climbing. So Deepak and Nielsen pushed at the pallet and lobbed it out like a howitzer shell. This was the pallet that landed furthest away from the target area, thanks to which the operation went awry. Nielsen and Deepak smiled at each other. Bleach called out to them, 'It was a struggle, but they are away now.'[20] They could not see the parachutes crack open to gently drift down the darkness over Purulia. The drop was, after all, not in the region of Panchet Hills near Dhanbad, as the 25 November 1995 R&AW intelligence indicated, but not too far away either.

As soon as the last pallet left the aircraft, the cargo door closed shut. Bleach gathered himself and returned to the main cargo cabin, which was a mess, with bits of box pieces, packaging material and other detritus. These he put into empty boxes, stacking them neatly at the rear of the aircraft. During the clean-up, two AK-47 rifles, which had dropped out of the nets, several magazines and some pistols were found. They were temporarily placed under a floor panel.

Soon after taking off from Varanasi's Babatpur Airport, Nielsen had instructed Klichine to turn off the cockpit voice recorder (CVR) which remained inactive for fifty-nine minutes.[21] By that time the AN-26 had flown over Gaya, where it deviated from the flight corridor. It was turned back on immediately before approaching Calcutta's Dum Dum Airport. There was no radio contact between the AN-26 and the air traffic control in Tarakeshwar, near Calcutta. Besides, no surveillance radar in the region was working at the time to detect the plane over Purulia. By the time the aircraft neared Calcutta, the cargo section looked perfectly clean and tidy. The remaining boxes were again covered with blankets and secured with straps, all luggage and moveables back in place. Everything looked normal. Bleach pulled out a novel, *Chains of Command* by Dale Brown, and began reading it.

To his utter dismay and annoyance, Nielsen learnt that Yangon had still not issued landing permission. Like the Bangladesh Civil Aviation Authority, the Myanmar military officials too seemed to have advance intelligence on the nature of the flight. That was when Nielsen instructed the pilots to land at Calcutta for refuelling. Klichine spoke to the Calcutta air traffic control tower and told Nielsen that the tower reported visibility of only 500 metres. The aircraft needed at least 800 metres of visibility to make a safe landing. Klichine wanted to brazen it out, saying that he would rather fly to Yangon, regardless of landing permission and, if necessary, call in for an emergency fuel stop. Nielsen acquiesced to this, so the aircraft overflew Calcutta, out over the Bay of Bengal. But suddenly, after about twenty minutes, Klichine turned the plane back, heading for Calcutta. Surprised, Bleach asked Nielsen what was wrong, and he said that the pilots had recalculated the fuel level, which now seemed insufficient for the AN-26 to reach Yangon. The return to Calcutta would be merely for a technical stop. The plane landed well past midnight at Calcutta's Dum Dum Airport, where Nielsen filed a fresh flight plan indicating Phuket as the next destination, with Surat Thani and Utaphao as two alternatives airports. Not long after, denied landing permission at Yangon and the aircraft's tank now refuelled, the flight took off for Phuket. Nielsen, Deepak and Bleach geared up to celebrate by the shores of the Thai sea resort, now that the mission had been accomplished. Or so they thought. They had combined courage and craziness which pulled them through their mission, but in the end, it was a botched-up operation.

7

NO RECEPTION COMMITTEE AT MUMBAI AIRPORT

On a cold and misty morning, Subhas Tantubai of Ganudih village woke up to take his cattle out grazing. It was a routine, boring chore for him. He brushed his teeth with a neem twig, splashed water on his face, wrapped his head in a grimy woollen muffler and draped himself in a frayed shawl to ward off the chill. Then, picking up the cowherd's stick, he set out with the cows to the fields. The grass was dry, so he decided to go beyond the open fields close to a rocky hillock where there were some saal trees and bushes for the animals to feed on. The sun had just peeped out, its rays piercing through the hanging mist and making visibility better. As the animals wandered about, Tantubai looked aimlessly here and there. Like his animals, Tantubai wandered off too. Suddenly his sleepy eyes caught the glint of metal near a grassy knoll by the hillock. Curious, he walked towards the object that had drawn his attention.

Tantubai's jaws dropped open. With cautious steps he moved towards the objects, picked one up, brought the cold steel close and then realized it was a type of gun that he had never seen before. There were at least thirty-forty of these scattered all over, the menacing muzzles protruding out of a mess of mangled wooden boxes. He thought for a moment and then, clutching the stick, ran as fast he could to Jhalda police station. His cows continued grazing.

Police at Bay

At the police station, Tantubai narrated his discovery to an on-duty assistant sub-inspector (ASI). He then ran to the Jhalda police station's officer-in-charge, Sub-Inspector Pranab Kumar Mitra's residence, where he narrated the discovery to the sentry on guard. The sentry knocked on Mitra's door. 'Sir, Subhas says he has just come across some weapons to the west of his house.' Mitra, who had just reached his official residence after a week-long leave and was sipping a cup of tea, dismissed the report, saying that the informant was perhaps still under the influence of country liquor. Seventeen years after 'those very tense but memorable days'[1] Mitra says,

> I continued sipping the hot tea when the Baglata circle inspector, who was a neighbour, hollered from his living room window and asked whether I was aware of the report. Now curious, I turned on the radio transmitter (RT) at 0700 hours. I could hear a lot of chatter about the discovery of huge quantities of arms and weapons. I quickly put on my uniform, wrapped a muffler around my ears, donned the khaki cap and drove to Chitmu village. The officer-in-charge of Joypur police station, Sub-Inspector Jiban Chakraborty, had already reached the spot. I saw some empty wooden olive-green crates that had been broken into and their contents emptied. The arms had disappeared.

Jiban Chakraborty had brought along an Indian army trooper, who lived in one of the villages. He asked him to dive into a pond in

Pranab Kumar Mitra.

Khatanga village to check if 'something could be found in the water', Mitra said, recollecting the chain of events that led to the discovery of the first weapon. 'As the soldier waded into the water, he bumped into something. He dived and pulled out an anti-tank grenade from the water. I was stunned. It was at that moment that we realized that this was a very serious matter indeed,' he said.

Mitra also spotted his superior officer, an additional superintendent of police (ASP), along with a deputy superintendent of police (DSP) and other lower-ranked officers and constables, desperately trying to locate the missing weapons. 'The Additional SP spoke into a mic, pleading with the villagers to return any weapon[s] that they might have taken away,' Mitra recalled. After taking stock of the situation, Mitra drove to the police station in his rickety jeep. He marshalled his men and wrote out an entry in the police general diary that he had stumbled upon some weapons which, as a policeman, he easily recognized to be AK-47 assault rifles. Excited, shouting orders to other subordinate officers, Mitra left the police station, driving around the nearby villages of Khatanga, Belamu, Maramu, Pagro and Beradih, where he found a number of Kalashnikov rifles scattered across several places, mostly on open fields and in thickets and trees. Someone reported sighting a huge piece of 'shiny, nylon-like' material (a parachute) under which lay more of the same weapons.

Mitra seized as many weapons as he could and screamed orders over an old walkie-talkie to other officers to spread out far and wide. Leaving his colleagues to recover more weapons, if there were any more to be found, Mitra returned to the police station, the third largest in the district extending 570 square kilometres,[2] gathered himself and made a few telephone calls to the Purulia DSP and the West Bengal State Intelligence Branch officer, among others. Mitra then filed a First Information Report (FIR), detailing what he and many of the villagers had seen, the seizures he had made and the number of village folks he had interrogated. Mitra, of course, had no clue how and from where the Kalashnikovs, rocket-propelled grenade launchers, anti-tank grenades and pistols, all of which he found to be very sophisticated, had landed. He had no time to rack his brains over the discoveries either. He simply wanted to be correct and appropriate in performing his duty in the face of a mystery. For Mitra, this was the case of a lifetime. In the case diary, he made copious notes of what he had seen in the villages and what he suspected was an 'act of subversion by agents inimical to the interests of the country'[3].

Mitra's description of the several places where the arms and ammunition were found appeared to be almost out of the discussions

that had taken place among the conspirators during the meeting in
Bangkok on 27 September, when Nielsen had spread out the survey
maps to point out the geographical features to Bleach, Randy, Mak
and Haestrup. At Belamu village, for instance, one of the two
parachutes was found in an open field. 'In the east is a vast open field
and in the west and north there are hills... At Ganudih, there are vast
open fields in the east and west... At Simnitar village, where the
second parachute was located, there is a jungle located in the southern
side of Pagro village and there is a vast stretch of land in the east and
west,'[4] Mitra noted, deeply puzzled by the shock discovery of such
sophisticated weapons in West Bengal's western-most district, shaped
like a funnel and girdled by the Tropic of Cancer.[5]

While Mitra and his officers got busy driving from one village to
another, seizing weapons and crates of ammunition, questioning the
residents, searching their homes and making detailed inventories of
the weapons recovered, people from neighbouring villages poured
into Jhalda police station to voluntarily hand over rifles and
ammunition that they chanced upon. There were also reports of some
villagers discovering the weapons and taking them away to their
homes as souvenirs.

The next day, 19 December 1995, the discovery of the mystery
weapons was all over the front pages of Calcutta's English and Bengali
newspapers. The newspapers reported the bare facts, quoting Purulia
district police and civil administration officers who were, needless to
say, clueless about the arms and ammunition. The Communist Party
of India (Marxist)-led Left Front government of West Bengal
panicked and took no time to inform the MHA which directed the
IB to find out what was really going on in Purulia. An unprecedented
event had occurred, which needed to be handled sensitively and in
absolute secrecy. IB sleuths landed in Purulia, the West Bengal police
top brass was camped in the district as the local police under the
leadership of a harrowed and helpless Mitra, along with Jiban
Chakraborty, continued with its search for missing weapons and
ammunition. The villages swarmed with cops in khaki and secret
service agents in mufti.

With the local police unable to make any headway, the West

Bengal government transferred the case to the state CID on 20 December 1995. Days went by and yet there were no answers. Where had the weapons come from? It was clear that an aircraft had used parachutes to drop the arms, but who could the conspirators be? Where had they flown in from? For days, there was speculation in government, bureaucratic, police and intelligence circles, but no department or agency could come up with sound and convincing answers. The CID, which was supposedly better equipped to solve difficult cases, failed to produce results too. These were early days and no breakthrough could be expected.

Revelry in Phuket

Far away on the idyllic white sand beaches of Phuket, the perpetrators made merry. After landing at the Thai sea resort, surrounded by the emerald green waters of the Gulf of Siam (now Thailand), those on board the AN-26 aircraft were met and serviced by Royal Thai ground support staff and driven to the terminal. All but Deepak

Deepak Manikan (left), Igor Timmerman (middle) and Niels Christian Nielsen by the beach in Phuket after the botched arms drop.

entered as aircrew; the tall, stylishly bearded Singaporean of Indian origin entered as a tourist. After the formalities were completed, the eight persons went by a small bus to Karon Beach Hotel, situated about an hour's drive from the airport. They all had separate rooms in the 'royal wing'. As soon as they checked in, the Latvians went for a swim. Bleach watched television before falling asleep. Nielsen and Deepak spoke about the drop and had a long conversation over telephone with Randy. After Bleach woke up, Nielsen, Deepak and he ate a late lunch around 1530 hours. By this time, of course, Nielsen knew the drop had gone horribly wrong. He said that a message had been passed from Randy to Hong Kong (possibly to Wai Hong Mak), saying that Randy knew where all the parachutes were on the ground, but one of them was very far apart from the others. Nielsen comforted himself, saying that this was not too alarming—it sounded as if Randy would recover at least the other two and the weapons' load was reasonably evenly divided.

After evening set in, the crew, Nielsen, Bleach and Deepak had a quiet dinner together. Nielsen and Deepak did not appear too worried that things had gone woefully wrong at the drop zone and took the Latvians to the nearest town, about fifteen minutes' drive from the hotel, to show them where the red-light district was. At the Karon Beach Hotel, there was a live band playing, which Bleach listened to, sipping a whisky. He was later joined by Igor Moskvitin and Evgueni Antimenko, who had not joined in the revelry of their fellow Latvians.

The next day, Tuesday, 19 December 1995, Nielsen had more bad news. He declared that Indian security forces were searching the drop zone and its adjoining areas, and stories on the massive quantities of weapons had appeared in the newspapers. Bleach advised Nielsen to call up Randy on the satellite phone to find out exactly what was going on. Nielsen remained silent and preoccupied. All he said was that much of the information was coming out of Hong Kong. It seemed that Randy and the insurgents he was in control of had not dared to venture out to collect the arms, as gun-toting security men were crawling all over Purulia. A few hours later, draft copies of choices of the new aircraft paint scheme arrived by fax from Hong Kong, which Nielsen showed to all. Everyone voted for the one which used a stylized heron in flight.

The remainder of the day was spent relaxing, some by the pool and some at the waterfront. Nielsen hired a jeep and two motorcycles that Deepak, he and some of the crew members used to drive around the area. Bleach stayed on at the beach, swimming and mulling over his next moves. In the evening, everyone gathered to have dinner together. They then went into town for some drinks at a local bar and to the market, where the crew bought clothing for themselves and their family members and Bleach bought some Christmas presents.

Fear Grips the Mastermind

On the morning of Wednesday, 20 December, there was still no clear information as to the situation in Purulia. Nielsen decided that Deepak should fly to Calcutta and try to get some first-hand information on the situation. Nielsen and Deepak went into town to get an air ticket, returning shortly after lunch to declare that they were unable to find an available direct flight to Calcutta, and so Deepak would fly to Singapore, from where he would take an outgoing flight to the West Bengal capital. Bleach was given the impression that Deepak would get the connecting flight almost immediately. Nielsen said that Deepak would have to leave the hotel by 1700 hours.

Around lunchtime, Bleach telephoned Syed Tasleem Hussain to have a long-overdue conversation with him in connection with problems relating to the 122-mm Howitzer ammunition contract. Bleach promised Hussain that as soon as he had sorted out the business at hand, which he explained was about the aircraft delivery and the proposed freight company, he would take the next available commercial flight to Dhaka to deal with the problems. Bleach said that he expected to be in Karachi on 21 December and would then be able to spend a couple of days in Dhaka before returning home for Christmas. When Hussain was pacified, Bleach sought to indirectly get information about the arms drop. He asked Hussain whether he had heard of rumours of terrorist activity in the Purulia region or read about it in the newspapers or seen anything on television. Hussain said he had not, but he could probably find an Indian newspaper and check. Bleach said that that would be very useful because the AN-26 aircraft would overfly the region and they did not want any trouble. Hussain said he would fax him any news.

As planned, Deepak left Karon Beach Hotel around 1700 hours. Bleach and Nielsen got together to find out what the flight schedule was to be like. Nielsen said he was considering flying to Indonesia, leaving the aircraft there and taking everybody by a commercial airliner to Karachi, where he would attend to re-registering the aircraft. The repainting could be done in Jakarta and the aircraft flown back to Karachi under a new registration. He seemed to have worked out the details of covering the tracks and creating the element of deniability. However, Bleach advised Nielsen that this would not work. His logic was that if the Indian authorities had by this time concluded that YL-LDB was involved in the arms drop, it would be quite simple for them to trace the aircraft, regardless of whether it was re-registered or not. He said that it would be a mistake to stay on in Phuket because the longer things were left unattended, the more time the Indian authorities would have to work out what had happened. His opinion was that it was still not too late to fly back through India. After all, there was a flight plan, not following which might attract more suspicion. The Indian authorities could have the aircraft seized in Phuket, a step which would not be very difficult to accomplish, at least as an interim measure. Furthermore, Bleach said, even if in the short term the Indian authorities went after Nielsen, the only really safe place for him would probably be Pakistan. Bleach's logic and argument sounded convincing to Nielsen. He agreed with the Briton's plan and it was decided to fly back to Karachi the next day, following the flight plan that had been filed earlier.

Later that evening, after dinner, Nielsen announced that he was going into town. Bleach said he would like to go along because he wanted to purchase a couple of items at the shopping plaza. They took a cab and Bleach asked Nielsen where the two should meet so as to return to the hotel together. Nielsen told Bleach not to worry and just take a cab back to the hotel, as he had to meet someone who had flown in from Hong Kong to deliver money in cash and a consignment of gold for a trial delivery by air to Varanasi, but had sent it back, fearing a search of the aircraft. Nielsen told Bleach that he would not be able to settle Bleach's dues when they reached Karachi, indicating that settlement would happen if and when they landed safe and sound in Europe.

The next morning, 21 December, Nielsen, Bleach and the crew arrived at the airport. The weather report predicted strong headwinds between Phuket and Calcutta. As the flight was on the edge of its endurance, it was felt necessary to look for an alternative plan. The weather report showed good tailwinds between Phuket and Madras, which meant they could fly to Karachi with only one fuel stop in Madras. Accordingly, an amended flight plan was filed. It was approved immediately and, shortly after lunch, the AN-26 took to the skies one last time.

The Last Flight

Once the plane was airborne, its rear was thoroughly cleaned. The last of the incriminating boxes and other trash were disposed off in the sea. The two Dragunov rifles, two night vision telescopic sights and three pistols (a fourth was discovered when Indian Customs searched the entire aircraft later in Mumbai) were hidden under the plane's floor. Nielsen had never intended to drop the rifles and telescopic sights by parachute over Purulia. He thought that they were probably too delicate and had planned to deliver them to Randy in Varanasi. He decided not to dispose of these weapons, so Bleach dismantled the rifles and Nielsen concealed them under the floor of the aircraft. The three pistols, together with some ammunition, were also concealed under the floor. It was planned that at a later stage the pistols would be licensed and form a part of the AN-26's standard equipment. The crew did not see Nielsen hide the rifles.

Soon the aircraft touched down at Madras airport without any difficulty. There were no security men and no 'reception party' to intercept the aircraft and its occupants. Everything seemed normal. When the plane came to a halt and was parked alongside other aircraft, Nielsen, Klichine and Moskvitin as usual reported to air traffic control to have the AN-26's papers stamped. Igor Timmerman attended to the refuelling procedure. The rest remained with the aircraft, which was on the ground for nearly two hours. Take-off was perfectly normal. Bleach stood in the cockpit during take-off, his pulse racing, worried that even at the last moment a message could come through the radio asking the pilot to return to Madras. He later negated the suggestion that Madras traffic control had ordered the plane not to take off and that it did so contrary to that instruction. 'This is absolutely incorrect,' Bleach wrote emphatically in his statement to the CBI.[6] Take-off was perfectly normal and there was no suggestion at any time of any impending problem. Yet, a fear of the unknown gnawed away at Nielsen and Bleach. They would not be safe as long as they were in Indian skies.

Bleach tried to sleep or read a book. Nielsen tried to catch up on some sleep too. Barely fifteen or twenty minutes from Mumbai, the cockpit radio crackled and the pilot was ordered to land at Mumbai airport. The cockpit door opened and Moskvitin, wearing his headset, beckoned Nielsen to join him. Nielsen went inside the cockpit, put on a spare headset and listened intently to a radio conversation. Nielsen's and Moskvitin's faces suggested that something was horribly wrong. So Bleach joined them at the cockpit door. He saw Klichine speaking into the mouthpiece while the others listened intently. Bleach could not comprehend what the problem was. Finally, when Nielsen removed the headset, Bleach asked him what was wrong. Nielsen, now appearing helpless, declared that the aircraft had been ordered to land at Mumbai. Klichine queried the order, a normal procedure, as Mumbai was not on the flight plan. The order was confirmed, stating only that the aircraft was required to land for investigation. The pilot acknowledged and altered course to make an approach towards Sahar International Airport. The remainder of his radio exchange with Mumbai air traffic control was technical and related to the landing. Bleach later claimed that at no time was the order to land disobeyed and, contrary to what was later suggested, air traffic control did not threaten to shoot down the plane if the pilot disobeyed instructions. During the pilot's conversation with air traffic control, Bleach checked out of all the windows to see if fighter jets had been sent up to escort the AN-26 down. He 'definitely' did not see any.[7] But the game was up.

Mumbai Airport

Once the aircraft began to lose height and the approach for landing started, Bleach joined Nielsen in the cargo compartment. Nielsen looked worried, but definitely not panic-stricken. He opened his briefcase, went through it in an orderly fashion and extracted a number of papers. These he tore up into small pieces and threw the shredded papers into a galvanized steel bucket in the aircraft. He then added a little bit of solvent and burnt the pieces. When the burning was completed, he added some water, turned the whole lot into a paste and flushed it down the toilet. Next, he took three or four floppy disks from the briefcase and broke them into small pieces. He borrowed Bleach's lighter, applied the flame to the internal disks and completely destroyed them. By the time this was finished, the plane's rear wheels had touched the runway.

Landing was normal and the AN-26 taxied to a parking slot behind a 'Follow Me' vehicle. Klichine switched off the engines. Nielsen and Bleach got out of the aircraft, which the Briton had expected to be surrounded by armed troops and a large contingent of officials. However, to everyone's great surprise, the area was completely deserted and engulfed in darkness. It was nearly 0140 hours when the aircraft landed at Sahar International Airport, Mumbai. The 'Follow Me' jeep had left and there was not a soul around. It took the entire lot, including Nielsen, a while to get over the surprise that none of the officials had found anything suspicious or even questioned them as to why the aircraft was there at all. It was simply unbelievable that the aircraft had been forced to land for being suspected to have airdropped the arms and yet there was no 'reception committee' on the ground waiting to greet it. They thought that the check must be a routine one, after all.

About ten minutes later an airport jeep, carrying two men, reached the aircraft. They identified themselves as the airport manager and his deputy and demanded to know what the aircraft's occupants were doing there and why the plane had made an unauthorized landing. They asked if the aircraft had had a fuel emergency. Nielsen volunteered to say that the aircraft did not need fuel; the pilot had been asked to land by air traffic control; he had no idea why but hoped that the officials might be able to shed some light. The airport officials said that the aircraft was poorly parked and the nose was partially blocking the taxiway. The pilot was asked to move it, so Klichine started up the engines, turned the aircraft and re-parked, this time to the officials' satisfaction. The two officials then called for a fuel tanker and an amused Nielsen and Bleach settled down to wait and see what would happen next. Nielsen and Bleach continued to talk to the two men, who clearly had no idea why the AN-26 had been asked to land. The stupidity and incompetence of Indian officialdom was on full display, with more to follow.

When Nielsen asked the officials whether they would have to pay landing fees, the senior of the two said he would have to. Nielsen, Bleach and the officials stood about 20 metres directly in front of the aircraft while the rest of the crew were either inside or stood near the plane. About fifteen minutes later (forty-five minutes after the plane had landed), another jeep arrived, carrying about six or seven men in plain clothes. They announced that they were Customs officials and wanted to search the aircraft. To Bleach and Nielsen, this served to reinforce the idea that the check would actually be routine. After all, the aircraft had flown in from Thailand, from where people often carried in contraband. Bleach thought that the Customs officials perhaps suspected the lot for drug smuggling. This was further reinforced by the attitude of the officials, who did not behave as if anything was particularly wrong. They too asked Nielsen why the AN-26 had landed, giving the impression that the search would be largely because the aircraft had made an unauthorized landing.

As the Customs men went inside the aircraft, Bleach immediately saw that there would be a problem, especially because the Latvian crew clearly did not seem to realize that the search party comprised Customs officers. So they started to object to the officials' entry into the aircraft. Bleach was anxious to avoid any problems, so he went to the aircraft and did his best to calm things down. He explained to the Latvians that the men were indeed Customs officers and that the crew must allow them to search anywhere they wanted. Meanwhile, Nielsen seemed to have sized things up and was with the airport manager. The fuel bowser (or tanker) had arrived after the Customs officials entered the aircraft. The AN-26 did not require extra fuel, so the refuelling process would not take long.

Nielsen's Houdini Act

About five minutes after the Customs officials entered the aircraft, Nielsen too got in. He picked up his document folder, which contained dollars in cash, and the aircraft documents, which he had always carried to air traffic control wherever the aircraft had previously landed. He calmly checked the contents, took some money out and left the aircraft. Bleach assumed that he would pay the landing fees and the fuel bill. He never saw Nielsen again. Bleach later learnt that before disappearing, Nielsen had handed over $2,000 to Igor Timmerman to cover the cost of the fuel. That was the money Bleach saw him remove from his folder before leaving the aircraft.

It took the Customs officials about forty-five minutes to search the aircraft and all the while Bleach remained with them. Klichine and Timmerman were also on the ground and spoke (mostly in hand gestures and some broken English) to some of the airport and Customs officials. It was when Bleach left with the Customs men that he saw that the aircraft had been surrounded by as many as fifty to seventy-five armed policemen. It was still very dark and he could only see their silhouettes. There was a large team of senior officers from a variety of different services and they all continued asking Bleach why the AN-26 had landed at Mumbai airport. The situation was confusing. Finally, after about fifteen to twenty minutes, one of the officers decided that it would be easier to sort things out in the terminal building. An officer came forward, demanding to see the passports of Bleach and the crew members. About ten minutes later, when Bleach had collected six passports, it became clear that Nielsen was not in the immediate vicinity. There were so many officers talking and shooting questions at the same time that Bleach simply assumed Nielsen had gone away with some group or the other and he would turn up later.

The six occupants of the AN-26 were taken to the airport building where Bleach was questioned by a number of senior officers from different services. It still seemed to him that the situation was confusing, to say the least, and no one really seemed to know why the aircraft had landed in Mumbai. The most repeated question posed to the seven men was why the aircraft had landed there. When Bleach said the pilot had only obeyed instructions, nobody seemed to believe him.

Even after an hour inside the terminal when Nielsen did not turn up, Bleach was not convinced that Nielsen had vanished and still thought it quite likely that he was being asked the same questions somewhere else in the same building. In any case, Bleach did not see why he should have disappeared at that stage. It was obvious from the moment the aircraft had landed that while Bleach, Nielsen and the crew members were being suspected of some wrongdoing, none of the officials appeared to be really sure whether the seven men had actually done anything illegal.

At some stage on Friday, 22 December, it became obvious that Nielsen had definitely and deliberately disappeared. The questioning turned to interrogation. Bleach was questioned by different officers, so he decided to wait until he was confronted by someone who seemed to know what he was talking about and why he and the crew had been detained. Only then would he tell the entire story. When senior officers took over his interrogation, Bleach spilled the beans and, as a result, the hidden guns were found on the aircraft. Bleach and the five Latvian crew members now faced arrest.

Phone Calls to Hong Kong

Subsequent investigation and questioning of some of the airport staff found (there will be more on the investigation in the next chapter) that Nielsen, who was carrying on his person no less than $20,000 in cash, probably took a ride on a jeep and hopped off at the terminal building from the apron[8] side of the airport, which is usually used for the entry and exit of VIPs, airport staff and air crew. Once a person gained entry there, 'there is no control on his movement and he can leave the airport building at will'[9]. There were two vehicles of the AAI and a 'Follow Me' jeep, also belonging to the AAI, at the spot where the AN-26 was finally parked after it landed. On the pretext of paying the landing charge, Nielsen had boarded the airport manager's jeep, which dropped him at the airport terminal building. It could be that Nielsen may have thrust a couple of hundred dollar bills into the airport manager's hands before disappearing into the terminal building. There is reference to Nielsen getting 'into a jeep' and disappearing in the report of the Parliamentary Committee on Government Assurances[10] which was supplied with the facts of the case by several Indian government ministries, departments and agencies, including the CBI.

The CBI found discrepancies in the statements of the airport and Customs officials, but zeroed in on the AAI officials, one of whom had offered to take Nielsen to the airport terminal building so he could pay the landing charge which, investigations later found, he never did. He could not have walked or slipped away from the parking bay without being noticed by the airport and Customs officials. After reaching the airport terminal building, Nielsen must have sensed he was free. All he had to do was to keep a straight face and then make himself scarce. But before leaving the airport terminal, he coolly walked over to a pay-phone booth and placed three calls, one of which was to an associate in Hong Kong (most likely his business associate 'King') which lasted no less than three minutes, and another to the suspected financier, Wai Hong Mak.[11] He spoke for half an hour to Mak (the call was recorded to have been made between 0419 hours and 0449 hours).[12] A third telephone call was made to a phone number in Jakarta, possibly to one of his Ananda Marga contacts based in the Indonesian capital. He then walked out of the old airport (Santa Cruz) building an hour before sunrise. Later that night, he called up the Pearl Continental Hotel in Karachi and spoke to one of the aircraft engineers, Alexander Lukin, instructing him to 'book tickets for the next flight to Riga' and followed up with another call late on the night of 23 December 1995, anxious to know whether the two Russian aircraft engineers had left Karachi.[13]

8

'YOU CONSPIRED TO WAGE WAR'

Setting off its rag-tag police force to locate the weapons and ammunition across several villages in Purulia had not produced substantive results for the West Bengal government. A large number of AK-47 rifles and tons of ammunition were still missing. Neither the state police nor its ill-equipped Intelligence Branch had any clue about the source of the weapons or the culprits who might have airdropped the arms consignment over Purulia. The IB had also begun making its own discrete inquiries and some of its senior and seasoned operatives had had several meetings with their counterparts in the R&AW, which had, as mentioned in an earlier chapter, provided early intelligence of a likely arms drop by a Russian cargo plane way back in the third week of November 1995. Director, IB, Dinesh Chandra Pathak, knew his organization had blundered, but then so had the MHA, the IB's controlling authority in government.

A day after the AN-26 landed in Mumbai, its crew was detained and some of the residual weapons and other equipment were seized, a desperate and edgy West Bengal police got cracking in Purulia. In a bold but knee-jerk action, the state police top brass, after consultations with the CPI(M) leadership, ordered a raid on the Ananda Nagar headquarters (at that time within the jurisdiction of the Jhalda police station in Purulia district) of the Ananda Marga.[1] On the morning of 23 December, 1995, at least thirty truckloads of policemen, armed with .303 rifles and batons, surrounded the ashram. The large posse barged into the sprawling complex after senior officers who led the raiding police party produced a court order. 'None of the top Ananda Marga monks (avadhoots) could be found. So we arrested eleven foreigners of various nationalities and produced them before a local court,' Sub-Inspector Pranab Kumar Mitra recalled.[2] Among the foreign nationals arrested was an American, David Ottee, described in a secret MHA Suspect Index as a 'violent and dangerous' individual 'requiring total surveillance'.[3] But within a couple of days they were out on bail, as no evidence of their involvement in the arms drop could be found. Irrational and inexplicable as the raid appeared at the time, it set the ground for later investigations by the CBI. Three months after the arms and ammunition were discovered, the CBI indicted four Ananda Marga monks (barring one, who, after being arrested in April 2004 in Delhi, passed away due to cancer in November the same year). Almost nineteen years after their names surfaced in the investigation, they continue to be 'absconders' and fugitives from the law.

On 26 December 1995, the septuagenarian West Bengal chief minister, Jyoti Basu, presided over a meeting of his cabinet, which took into account the extraordinary situation arising out of the arms drop and the threat to not just the state's but the country's internal security. In view of the 'gravity of the situation', West Bengal's Marxist government took the collective decision to hand over the mysterious case to the CBI by an executive order. The CBI's Special Crime Branch in Calcutta took up the case on 27 December 1995. The young CBI SP, Loknath Behera, directed a DSP, P.S. Mukhopadhyay, to take up the formidable investigative challenge. Assigned as the investigating officer (IO) for the case, Mukhopadhyay left for Purulia the following day. After long consultations with the officers of Jhalda police station, he prepared the criminal case, which he duly numbered Regular Case/SCB/11/95-Cal. The central government had set a lumbering elephant after a gazelle.

Investigations Begin

Mukhopadhyay's first task was to thoroughly read the FIR which Pranab Kumar Mitra had filed on 18 December 1995. As the procedural matters of the CBI case fell into place, Behera, in consultation with his joint director at the agency's New Delhi headquaters, J.K. Dutt, dispatched a team of investigators to Mumbai to 'seize' the AN-26, arrest the six crew members, including Peter Bleach, and bring them over to Calcutta on a transit remand for interrogation along with some of the weapons and equipment, including the GPS and a laptop that Nielsen had left behind. A second, much larger team of officers, this time headed by Behera, reached Jhalda and, over the next two weeks, collected all of the weapons and the crates-full of ammunition, before bringing them over to be stored in a sprawling room at the West Bengal state police lines in Barrackpore, North 24 Parganas district.

Part of the weapons cache stored at the police lines in Barrackpore, North 24 Parganas, West Bengal.

Meanwhile, *The Telegraph* published the first in a series of investigative stories on the Purulia arms drop, providing fairly accurate accounts of Bleach's meeting with Nielsen, Peter Haestrup and Brian Thune in Copenhagen in the middle of August 1995; the 27 September 1995 Bangkok conclave of the conspirators where Purulia was decided as the arms drop site; the actual purchase of the aircraft in Riga; the loading of the arms and ammunition at Burgas Airport in Bulgaria; Nielsen's gold smuggling ventures in India, Dubai, Thailand and Hong Kong, besides his 'unlimited cash reserves'; the presence of

Randy at the Bangkok meeting; and the investigators' initial suspicions that the weapons were probably intended for the Ananda Marga.[4]

Among some of the Mumbai airport officials who were examined by the CBI towards the end of December 1995 and in the first week of January 1996 was one Zainulabidin Abdul Karim Surve, who was the airport manager on duty at Apron Control II from 2000 hours of 21 December to 0800 hours of 22 December. Around 0130 hours on 22 December, when he was at the office of the senior airport manager, Surve received a message from the security supervisor, S.T. Jani, that a non-scheduled aircraft had been directed to land in Mumbai and its expected time of arrival was 0139 hours. He claimed that he 'immediately informed the senior manager, A.K. Puri, regarding the arrival of the non-scheduled aircraft and came to Apron Control II around 0140 hours'.

According to Surve,

> Mr Dasan, an IB officer, was inquiring about the non-scheduled flight landing from Mr Jani, who furnished the required information to him... Around 0145 hours I left Apron Control II on [a] jeep (No. MH 02K-1043) driven by the driver, Mr R. G. Jaiswar... When I reached near the aircraft around 0215 hours, I found that it was already parked. Mr Balbir Singh [the assistant airport manager] and the security supervisor, Mr Arun Mukar Tirkey, were standing near the 'Follow Me' jeep and I also noticed a jeep of Hindustan Petroleum parked near the aircraft. Three crew members were also on the ground.[5]

Surve saw that one of the men on the flight (Bleach) was in a white safari suit and another was comparatively shorter in height, with a beard and wearing dark trousers and a coloured shirt (Nielsen). He claimed that around 0230 hours he sent Arun Tirkey to Bay No. 10, where a training programme was to be conducted, and asked him to take the jeep driven by R.G. Jaiswar. Tirkey had to be sent because he carried the radio transmitter that was required at the training programme.

It was Surve's contention that once the jeep left, a white Maruti

Gypsy reached the spot carrying four to five persons (Customs officials). 'Mr Dasan of IB was one among them,' Surve said, adding that soon, a search inside the aircraft began. 'He left the site after 0300 hours for Bay No. 10 to supervise the training programme...When I reached the old airport, police informed us that one person was missing from the aircraft...

'Here I want to make clear that no outsider travelled in my jeep or in the jeep along with Mr Tirkey on the night in question. I have no idea as to how Mr Kim P. Davy (the name I came to know subsequently), the man with the beard, escaped from the scene,' Surve said. Surve claimed the last he saw Nielsen was around 0220 hours.

But when Customs superintendent (Airport Intelligence Unit), Amirali Chagan Bawa, was questioned by the CBI, he said that when he reached the aircraft around 0245 hours along with a few of his colleagues and IB's Dasan, he saw 'one Airport[s] Authority of India jeep and a "Follow Me" jeep as well as an HP (Hindustan Petroleum) jeep and an NR5 oil tanker'[6]. He found three crew members (Nielsen, Davy and Timmerman) standing near the aircraft. Bawa said in his statement that after the search operation, when all the Customs officials and IB's Dasan emerged from the aircraft, 'we did not find either the AAI vehicle or the AAI officials. By that time, the aircraft had been surrounded by armed policemen.'[7]

In his statement to the CBI, Hindustan Petroleum official R. Jyoti Prasad said that soon after he reached the aircraft at 0215 hours, he saw AAI staff trying to help in the process of parking the AN-26 'properly'. Even at that time, Prasad noticed 'two crew members. One person was more than 6 feet tall, lean, fair and was wearing a white safari suit [Bleach]. Another person was of medium height, medium build, fair and was sporting a beard [Nielsen]. He was wearing dark trouser [sic] and a coloured shirt'[8]. According to Prasad, a second AAI jeep 'arrived near the aircraft' soon after he reached the spot and that 'one jeep of [the] AAI left the scene just before the aviation fuel bowser reached the place.'[9]

From the statements it can be easily deduced that Nielsen escaped from where the AN-26 was parked before the aircraft was surrounded by armed policemen. Neither Bleach (so he claims) nor any of the crew members noticed whom Nielsen spoke to last before he disappeared. It seems also that none of the officials near the aircraft kept an eye on the 'Follow Me' jeep. Both Surve and Balbir Singh claimed that the two AAI jeeps left the spot without any other person on them. Both could have lied to save their skins. Nielsen couldn't have hitched a ride on the jeep that Tirkey took to reach Bay No. 10 because it was being driven by Jaiswar, who surely would have noticed another person climbing onto the jeep.

It is certainly unlikely that Nielsen walked away from the aircraft or slipped into the darkness without any of the officials present at the time noticing him. He couldn't have walked the huge distance from the aircraft to the airport perimeter wall without being challenged by security men. The only plausible course of action on his part would have been to approach the airport terminal building, which he could have done by taking a ride on a vehicle. Besides, as mentioned, there is evidence to indicate that he made three telephone calls from a pay-phone booth, which could only be available to him at the terminal building, a crucial point that connects his disappearance from the spot where the aircraft was parked to a ride on a jeep and his emergence at the building.

Although the CBI initially took down the witness statements of Surve, Balbir Singh and Tirkey, the three AAI officials were later subjected to intense interrogation in which it was found that they had lied. It was discovered that when Nielsen had inquired of Surve whether landing charges were required to be paid, the airport manager had confirmed it. Nielsen then went inside the aircraft, formatted the hard drive of his laptop, handed over $2,000 to Timmerman to pay for fuel ($600 was subsequently paid for slightly over 2,000 litres of aviation fuel) and joined Surve in his 'Follow Me' jeep, ostensibly to pay the landing charges. He walked into the building but never paid the landing fee.

The then director, IB, Dinesh Chandra Pathak's response to a CBI questionnaire on the sequence of events that led to Nielsen's escape from Mumbai airport was not just a comedy of errors, it exemplified how utterly callous, negligent, incompetent and what an accomplished liar the chief of the country's internal security agency was.

I recollect that on December 22 at about 0045 hours, I received a call from AVM V. G. Kumar, Director, Air Intelligence, informing me that the suspect aircraft had landed at Madras Airport and seeking assistance to see that it was given any clearance to take off. I immediately contacted DD [Deputy Director], SIB [Subsidiary Intelligence Bureau], Madras, and directed him to alert the immigration authorities at the airport about the suspected aircraft and to ensure that it was not allowed to leave. DD, SIB Madras, after checking up with the airport, informed ne [sic] that the aircraft had in fact landed at Madras at 2045 hours on December 21 and had already departed at 2245 hours for Karachi.[10]

Pathak was not even aware of the whereabouts of the AN-26 on 22 December 1995.

After receiving a second phone call from A.V.M. Kumar the same night, Pathak claimed he

...never left the telephone that night since I was woken up by the first call of AVM Kumar. My priorities were to instruct the head of SIB, Mumbai, to get the immigration into full swing and to have the Mumbai Police mobilised. Immigration is in direct command of IB and, [the] DIB, though not a 'force commander', himself is in a position to operationally advise the State Police, which I did by reaching out to [the] Commissioner of Police, Mumbai, without any avoidable loss of time.[11]

The CBI, not satisfied with Pathak's response, found him guilty for failing to take appropriate timely action, including not alerting the immigration authorities in Mumbai. Besides, in his response to questions by the CBI, then Mumbai Police Commissioner R.D. Tyagi said he was informed by the DIB about the suspect aircraft only around 0235-0240 hours, i.e. two hours after Pathak received the information and an hour after the aircraft had actually landed (0139 hours) at Mumbai airport.[12] The CBI's investigation also revealed that Pathak had called up his subordinate in Mumbai, Deputy Director

C.M. Ravindran, between 0130 hours and 0140 hours, i.e. an hour after the initial receipt of information from AVM Kumar. A more authentic account of the events that followed after the AN-26 was 'force-landed' at Mumbai airport was reflected in a Special Bureau (the main R&AW station in the Maharashtra capital) memo of 22 January 1996. It said, inter alia:

> It seems that the officer-on-duty from MLU informed ATC [air traffic control] at 0040 hours that the flight 'YLLDB' was to be landed for 'investigation.' The officers manning [the] ATC asked for instructions, in writing, so the MLU representative informed this in writing. The aircraft was force-landed at 0139 hours and one MLU officer was present in the ATC tower at the time of landing. Within a few minutes, [the] ATC also received a call from [the] immigration authorities enquiring whether they had succeeded in landing the aircraft. At 0215 hours, the IB informed Customs about the landing and requested for rummaging of the aircraft. At 0240 hours, the Police Commissioner, Mumbai, got a call from DIB informing about the landing. At 0255 hours, [the] ATC informed the airport police about the landing of the aircraft.[13]

What was found alarming in the course of the CBI's investigation was that the IAF, which force-landed the AN-26 at Mumbai airport, did not inform airport security, the police or the immigration authorities about the suspect plane, leading to utter confusion, compounded by the IAF's stand that since the aircraft had landed at the civil airport, the IAF had nothing to do with the security arrangements.

A subsequent meeting of the Maharashtra Anti-Terrorism Group in Mumbai concluded, as surmised above, that even as there was total confusion among airport, security and Customs officials, Nielsen appeared to have convinced one of the AAI staff near the aircraft that he had to go the terminal building airside to pay the landing charges. '...the man was escorted by someone from the AAI,' the minutes of the meeting of the Anti-Terrorist Group noted.[14] Clearly, a lackadaisical attitude on the part of the AAI and security officials, who had no idea or did not comprehend the gravity of the situation, contributed in no small measure to Nielsen's escape from Mumbai airport.

Taking full responsibility for the disastrous events at Mumbai airport, the then union home secretary, K. Padmanabhaiah, later admitted that the MHA had failed to coordinate security arrangements.

If there was lethargy and lack of coordination among airport and security staff—if not plain dereliction of duty—at Mumbai airport from where Nielsen escaped with ease, there were glaring procedural holes at Madras airport where the AN-26 landed and then took off for Karachi without even the mandatory ADC that was supposed to have been issued by the IAF for a foreign aircraft landing and overflying India. The bungling by Indian authorities, first by allowing the aircraft to fly to Madras without an ADC (while it was on its way from Phuket to Karachi) and then failing to prevent the plane to take off from Madras, took place at a time when the story about the arms drop had already gone viral in the Indian print and electronic media and the entire world was aware[15] that huge quantities of arms and ammunition had been airdropped over Purulia three days ago. When the IAF was alerted that an unscheduled foreign aircraft was suspected to have airdropped the weapons, a team of IAF officers, including a paratrooper from the air force station in Agra, reached Purulia on 21 December 1995, and made a positive identification of the YL-LDB after checking the entry and exit records of all unscheduled flights that had operated between 18 and 21 December. That information was shared with Madras airport late on the night of December 21, only after the aircraft had already taken off for Karachi.[16] The parachutes were identified to have been of South African make. According to Bleach's statement to the CBI, the AN-26 landed in Madras on schedule without any difficulty and remained on the ground for a full two hours before it took off for Karachi.[17]

The CBI's investigation found that for the Phuket–Madras sector, the DGCA did not seek an ADC from the crew of the AN-26 aircraft and, therefore, it was not issued by the IAF. The aircraft landed at Madras airport without an ADC, but the IAF's military liaison unit there granted the ADC for the flight between Madras and Karachi! In his effort to connect all the dots and prepare a sound and foolproof case, CBI SP Loknath Behera directed the agency's Madras-based investigators to examine and record the statements of air traffic control and other officials who were on duty at Meenambakkam Airport on the evening of 21 December 1995, when the flight landed.

The Madras Flight Information Centre (FIC) aerodrome officer, T. Tamil Selvan, who was on duty between 1300 hours and 2000 hours on 21 December 1995, admitted to receiving a flight plan from Phuket in respect of AN-26 around 1430 hours:

> As per this flight plan, the AN-26 was to land at Madras (Chennai) for refuelling on its way to Sharjah via Karachi. Thereafter, a series of information was received from various flight levels into [sic] facilitate landing of the aircraft, to clear air space and parking etc... We do not keep any watch as to whether all the authorities of DGCA are received by FIC, since it does not serve any purpose.[18]

He added that no information in respect of the proposed landing of the AN-26 aircraft was passed on to the briefing section in advance because 'FIC receives information from other airports only for the purpose of clearing air space, communication with incoming aircraft, parking facilities etc. Even on earlier occasions, in respect of other flights no information was passed on to briefing'[19].

During his examination, the senior aerodrome officer at Madras airport at the time, S.G. Sathe, who was attached with the briefing section—whose responsibility was to accept flight plans before passing them on to the control tower and the area control centre—said that usually flight plans are received by the briefing section once they are cleared by airport security, immigration, Customs, the health department and the MLU. A flight is allowed to take off when it is cleared by the briefing section. Sathe informed the CBI interrogators that when a flight plan is received under normal circumstances, its authority is verified with the check register. In case an authority is not entered in the register, the DGCA is contacted to confirm the authority. Besides, flight authorities from the DGCA are received by the MLU too.[20]

In the case of [the] AN-26, the flight plan and general declaration were received by [the] briefing [section] by 2200 hours. When it was cross-checked with [the] register, no entry was found for flight authority. Since the watch supervisory officer was busy with his work at radar control, when an attempt was made to take it to his attention and since it was cleared by all wings, including MLU and in view of the fact that DGCA could not be contacted during odd hours, clearance was given to the AN-26. A flight could not be detained without valid reasons. Briefing had only very little time to process the flight plan.[21]

The senior aerodrome officer added that 'the message of the proposed landing of [the] AN-26 was received by FIC by 1430 hours. However, it [the flight plan details] was not sent to [the] briefing [section].'[22]

No airport official, including those of the AAI or the IAF's Madras MLU, suspected why the aircraft, which was originally supposed to re-enter Indian airspace and fly over Calcutta on its way to Karachi, suddenly changed course and decided to fly to Karachi via Madras. They believed the pilot's explanation (who was evidently tutored by Nielsen) that the AN-26 had to change course because of poor weather conditions in Calcutta. In his response to the Parliamentary Committee on Government Assurances, the then director general of civil aviation, H.S. Khola, said there was nothing unusual about the aircraft landing in Madras even though its original flight plan was to fly to Karachi via Calcutta. 'There was no reason to question the aircraft as it had the permission to land in India,' Khola told the committee.[23]

Customs officials failed to perform their duty at Madras airport after the AN-26 landed. The mandatory submission of general declaration, which included filing the passenger and cargo manifests as well as details of the crew members' private property, was not done. In fact, one Customs officer said that 'entries in respect of four or five non-scheduled flights that landed at Madras during December [1995] were not made…No non-schedule flight was inspected during the month of December, 1995.'[24] Lapses on the part of airport security were also recorded by the CBI. According to Madras airport security inspector, V. Pattaviraman, while he did visit the aircraft at the parking bay and verified that there were seven crew members, he 'did not enter [the] aircraft'. Pattaviraman declared that 'there is no practice of obtaining a copy of the declaration with a security seal' duly signed by the officer giving clearance. 'Similarly, no register is maintained for having accorded clearance to (various) aircraft.'[25]

When the AN-26 landed at Varanasi late in the evening on 17 December, there was no way Indian officials could know that the aircraft's lethal cargo was being fastened to the parachutes that Deepak had brought into Karachi from Johannesburg. This had caused a delay of over three hours before the aircraft took off for Calcutta. The delay was attributed to 'technical reasons' and conveyed accordingly to air traffic control officials at Varanasi's Babatpur Airport. According to a flight plan filed at Babatpur Airport, the AN-26 was to land at Dum Dum Airport in Calcutta[26] on its way to Yangon, where Burmese officials had not yet given permission to the aircraft to land.

Subsequent investigations found that no ADC was sought by the civil aviation authorities for the aircraft's flight between Varanasi and Calcutta. Even though the aircraft was at Varanasi airport for over four hours (it landed at 1735 hours and took off for Calcutta at 2200 hours on 17 December 1995), no Customs check was conducted. The Customs superintendent, Om Prakash, who was on duty at Babatpur Airport that evening, gave three reasons for not conducting the check— the aircraft was of foreign origin, was in transit and had landed for refuelling; no loading and unloading of goods was done; and there was neither any reasonable belief nor any suspicion that the aircraft, which had landed for refuelling, was being used for smuggling in any contraband/prohibited items.[27] Besides, since the AN-26 declared that it was a 'ferry aircraft' (meaning it was not carrying any goods), Customs and security officials did not perform any checks on it. During the November 1995 'dry run', when the AN-26 remained parked at Babatpur Airport, Nielsen had observed that the airport staff and security personnel were slack in performing their duties. He had had 'free access' to the airport building, which 'facilitated' his plans for a subsequent visit with the weapons.[28]

However, what came to the notice of the CBI was that the DGCA did not seek an ADC from the IAF for the aircraft flying between Varanasi and Calcutta. In responding to the CBI's queries, the civil aviation ministry sought to take the plea that in the case of the AN-26, clearance was granted when it flew into Indian airspace after it took off from Karachi. The then AAI director (operations), P.C. Goel, tried to squirm out of the situation, claiming that the ADC obtained from the MLU in Mumbai was supposed to be valid right up to Yangon, but he could not come up with a satisfactory reply when questioned about why no IAF clearance was sought even when the aircraft halted at Varanasi airport for more than half an hour. Apparently, the Varanasi air traffic control transmitted the AN-26's flight plan to Dum Dum Airport in Calcutta, where airport officials transmitted it to the IAF MLU, which had the authority to amend or validate an ADC (in this case, the one that was obtained when the aircraft entered Indian airspace). The CBI later discovered that the DGCA did not seek any ADC for the AN-26 even when it landed at Varanasi airport as part of a 'dry run' before the actual arms drop, on 27 November 1995.

Once it was impounded, a search of the AN-26 yielded a treasure trough of information that the CBI would later use in a Calcutta court of law as documentary evidence against Peter Bleach, the Latvian crew members, Niels Christian Nielsen and four Ananda Marga monks for their alleged involvement in the arms drop. Among some of the important material the CBI found was a laptop that belonged to Nielsen, his briefcase containing vital papers related to the purchase of the AN-26 and the arms and ammunition, a GPS, several letters he had received from his associates in Denmark and Hong Kong, photographs of the arms drop zone and of an imposing 'white building' in Bansgarh mouza[29] (in the jurisdiction of Jhalda police station), and diagrammatic sketches of the parachutes rigged to the crates containing the weapons and ammunition. Some of the documents clearly indicated the places where the aircraft had landed and refuelled on its way to India—hotels and refuelling bills and receipts of payments made at various airports in Bulgaria (Plovdiv and Burgas), Iran (Zahedan and Espahan), Pakistan (Karachi), India (Varanasi and

Calcutta) and Phuket. In Nielsen's stylish Franzen leather briefcase, investigators found a birth certificate (No. 680724) issued at Rotorua, New Zealand, on 21 February 1991 in the name of Kim Palgrave Davy (the first indication that the name Kim Palgrave/Peter Davy was fictitious), two different driving licences issued in the name of Kim Palgrave Davy in the Philippines, and much more.

On the face of it, the wealth of information suggested that it would take no time for the CBI to identify the perpetrators behind the arms drop and 'crack' the case, but SP Loknath Behera was fully aware that it would be a long haul. He was confident that while the documents recovered from the aircraft and Nielsen's briefcase would be invaluable, part of the mystery could be unravelled if the laptop could somehow be 'broken into'. He promptly sent the laptop to the Hyderabad-based Central Forensics Science Laboratory, where it took software experts about two months to retrieve all the data in the machine's hard drive. Behera was satisfied with the results achieved so far. Not only did he now have printouts of emails Nielsen had exchanged with Bleach, Peter Haestrup and Brian Thune, he also accessed electronic correspondence exchanged between Nielsen and some Ananda Marga monks, including the then secretary general of Proutist Universal, an important wing of the Ananda Marga cult, Acharya Tadbhavananda Avadhoot.

The Ananda Marga Monks

For Behera, a blurred picture had begun to emerge about Nielsen's antecedents, his spiritual inclinations and his links and connections with the Ananda Marga. When communicating via email with Ananda Marga monks, Nielsen would use the email ID 'DNA'. On other occasions, when writing to Bleach, Haestrup and Thune, he would use the ID linked to the assumed name Kim P. Davy.

Loknath Behera.

The emails to Bleach were a clear indication for Behera that there must be other paper trails. From the emails exchanged between DNA and the Ananda Marga monks, Behera got the impression that something sinister was afoot. Some of the emails referred to 'arms training' and 'defeating the commies'. A weak but important link between the arms that had been airdropped over Purulia, very close to the Ananda Nagar headquarters of the Ananda Marga, had begun to come to the fore. However,more evidence was required.

Earlier, towards the end of December 1995, CBI officers, like their West Bengal Police counterparts, raided and searched the Ananda Marga's Ananda Nagar ashram, and in one fell swoop carried away masses of documents, books, letters, copies of speeches, plans and programmes in print, cameras, photographs, video recorders and computers, and other peripherals. Raids and search operations were also carried out at several other Ananda Marga establishments in Calcutta and Bangalore (now Bengaluru). During the raid and later, the CBI recorded the statements of villagers living in the vicinity of Ananda Nagar who claimed to have seen and heard some Ananda Marga monks, including Acharya Tadbhavananda, Acharya Jagdishwarananda, Acharya Saileshwarananda and Acharya Suranjanananda (Randy, alias Satyendra Singh, alias Satyanarayana Gowda), speaking to followers about self-defence and arms training.

For instance, Kartik Pramanik, a resident of Garudih under Jhalda police station, described the Ananda Margis as a 'very violent type of people' some of whom, including Saileshwarananda and Vinay Singh (Randy's younger brother, subsequently arrested and chargesheeted by the CBI for his alleged role in the arms drop), 'fired [at] and killed villagers' without provocation. Pramanik alleged that 'a few years ago, Vinay, the brother of Suranjanananda, once fired from his gun and injured a villager'. According to Pramanik, before the discovery of the arms and ammunition in Garudih and other adjoining villages, he would often see Saileshwarananda and Vinay Singh, active members of the VSS (Voluntary Security Service), together at Bansgarh. However, 'since December 18, 1995, both of them have disappeared'.[30]

Montu Gorai, also of Garudih, made an identical statement, but claimed that Randy, who was second-in-command of the VSS after

Saileshwarananda, 'is close to Pappu Yadav, MP from Bihar', a point which never came up during the trial of the Purulia arms drop case, but would assume great importance in 2011 (more about Pappu Yadav's alleged links with Nielsen later). According to Gorai, who was employed as a driver by the Ananda Marga between 1985 and 1990, Nielsen would often visit the Ananda Nagar ashram, 'especially during the Dharma Maha Sammelan (large religious congregation).'[31]

A more revealing statement was made by Barun Chandra Kumar of Baro Rolla village. Kumar alleged that 'in the first week of November 1995, a training camp was organised by the Ananda Margis in Ananda Nagar'. An Ananda Marga monk, who Kumar identified as Acharya Raghabananda Avadhoot, tried to convince him to join the camp, like others from his village. On a particular day (he did not mention the date), 'on arriving at Ananda Nagar, I could see that training was on. About 80-85 women and about 100 men were being given training on rifle shooting. The trainer's height was about six feet, fair complexioned, [he] wore half pants, a khaki shirt and [had] a gun in his hand,' Kumar said, identifying the 'trainer' as Vinay Singh.[32]

Referring to the two-storeyed 'white building' in Bansgarh, Paritosh Saha of Taherbera village alleged that 'regular training' was held for about two to three months around the house prior to the arms drop of 17 December 1995. That evening about '10-15 Ananda Margis' had assembled at the house of Dhananjay Mahali who taught in an Ananda Marga-run school,[33] Saha claimed, recognizing Randy and Vinay Singh from four photographs shown to him by a CBI officer. The other two photographs were of Nielsen and Deepak.

In their statements to the CBI, several others villagers claimed to have seen Saileshwarananda and Tadbhavananda at VSS training sessions. One resident of Bhagudih village, Riju Majhi, alleged to have seen Nielsen (he recognized him from a photograph shown on a television news programme after the arms drop) in the company of Tadbhavananda near the Bansgarh building. 'That time they were carrying camera etc,' Majhi alleged.[34]

An additional, related piece of information that the CBI could lay its hands on appeared to corroborate the allegations made by some of the villagers that the plot of land on which the two-storeyed white building stood had been illegally occupied by the Ananda Marga.[35] A document, issued by the Purulia District Land and Land Reforms Settlement Officer on 8 December 1996, inter alia said:

> It appears from our record of right finally framed and finally published under the Provision of West Bengal Estate Acquisition Act, 1953, that there was no building over the said plot. [The] Government of West Bengal is the owner of the said land since 1985.
>
> However, it appears from our record-of-right finally framed and finally published under the provision of the West Bengal Land Reforms Act, 1955, that a building is located on a portion of R.S. plot No. 210 measuring a total area of 19.98 acres of the said *mouza*. Ownership of the land is vested with the Government of West Bengal since the declaration of vesting in the year 1970 vide B.R. Case No. 64/1970 under the provision of Section 6 of the West Bengal Estate Acquisition Act, 1953.
>
> Prior to the declaration of vesting, the land was held by one Chandra Mohan Tewari, S/o (son of) Shital Chandra Tewari of village Bansgarh.
>
> Some portion of R.S. plot No. 210 is unathorisedly (sic) occupied by the Ananda Marga Pracharak Sangha as it is revealed from the relevant record-of-rights.[36]

Nine months before the CBI stumbled on this revelation, it had filed (within the stipulated ninety days from the lodging of the FIR) in the court of the Ninth Metropolitan Magistrate, Bankshal Court, a lengthy chargesheet against thirteen persons (Kim Peter Davy, alias Niels Christian Nielsen; Peter Bleach; Deepak Manikan, alias Daya M. Anand; the five Latvian crew members; the four Ananda Marga monks; and Rameshan Bhanu, of Alleppey district of Kerala, whose involvement in the arms drop was not clear), under various provisions of the Indian Penal Code (IPC), accusing them of conspiring and attempting to wage war against the state. The maximum punishment prescribed by the IPC for crimes committed under the provisions of the act was death. A couple of months after the first chargesheet was filed on 20 March 1996, the CBI arrested Vinay Singh, who had disappeared from Ananda Nagar since the arms drop, from somewhere in Purulia. A supplementary chargesheet was filed against Singh on 5 September 1996 in the court of the Ninth Metropolitan Magistrate in Calcutta before the case went in for trial on 1 August 1997. In the three months preceding the issuance of the first chargesheet when the CBI battled a deadline, the investigation continued at full pitch. It had been all too hasty, too ill-planned, leaving gaping holes in the evidence collected and their logic.

The CBI's case against the Ananda Margis—or at least a section of the monks—centred around the two-storeyed white building which 'could be ascertained that it belonged to the Ananda Marga organisation and was being used by them for the purpose of training V.S.S. members... The building was searched and among other articles a packet [containing] some powder, which was tested as an explosive substance by experts of the Central Forensics Science Laboratory experts [sic], was seized.'[37] The CBI then discovered Randy's presence at the 27 September 1995 Bangkok meeting of the conspirators where, among other things, Nielsen produced survey maps to indicate the drop zone as well as photographs of the white building (the supposed drop site). The maps and photographs were found in his briefcase that he had left behind in the aircraft before escaping from Mumbai airport.

Retrieval of data from the GPS after analysis by the IAF's Directorate of Intelligence revealed that the crew (obviously on the direction of Nielsen) had fed the longitude and latitude coordinates of as many as forty-nine 'waypoints' over which the AN-26 intended to fly before it took off from Burgas Airport in Bulgaria. The chosen heights above these waypoints were also fed into the GPS to indicate to the pilot, Alexander Klichine, the exact location over which the aircraft would fly. Two of the forty-nine waypoints fed into the GPS—Sarek and Taget—drew the IAF's attention. They were not found to be reporting points and were not part of the air traffic route. Among all the coordinates, one—23° 28'45" North and 86° 01'56" East (see Chapter 1, R&AW's intelligence report)—was programmed into the GPS manually on 15 December 1995, whose corresponding height was found to be 1,500 feet.[38] 'The heights planned [for the two waypoints] were very low and are ideally suitable for para-dropping. Civil aircraft are not permitted to descend to such low levels in these areas. All air traffic routes start at 6,000 feet and above. Waypoint 37 (Taget) has the height of 500 feet and waypoint 42 (Sarek) has been given 1,500 feet. This indicates that the user wants the GPS to indicate these places when the aircraft is 500 feet and 1,500 feet above the respective places,' the IAF's analysis concluded.[39]

An analysis of the second waypoint's coordinates by the Survey of India described the waypoint to be in Purulia district. 'This is a place at the foot of the NE (northeast) edge of hill which is a protected forest JHORA-PAHAR on the Bihar Bengal border... The point is situated in Bansgarh village, Purulia district, West Bengal,' the Survey of India report revealed.[40]

For the CBI, this was clinching evidence not just indicating precisely that Bansgarh was the drop zone, but a means to link the Ananda Marga, or at least a small section of the organization's monks, as the recipients of the arms and ammunition.

More ominously, the IAF's analysis of the data from the cockpit voice recorder (CVR), the flight data recorder (FDR), and the velocity height gravity recorder (VHGR) revealed that:

1. The data shown in the analysis report of FDR pertains to a total duration of 59 minutes. However, exact date and time cannot be determined.

2. The FDR analysis report does not correspond/tally with the analysis report of the CVR and VHGR. The following are the glaring contradictions:

   a) FDR recording has been only for a duration of 59 minutes. However, VHGR recordings are available for total six sorties/flights. The CVR recordings are available for a duration of total (the rest is not legible)

   b) The last records of the VHGR, CVR and FDR do not tally with each other.

   c) CVR recordings sometimes lead us to conclude about a particular date and time. However, VHGR/FDR recordings does not record any date and time.

3. The CVR recordings do not tally with the known flight plan of AN-26 (YLLDB) on 17th Dec 95 and 18th Dec 95 i.e. from Varanasi to Calcutta. The CVR and VHGR records do not indicate date and time and hence could belong to any flight. Unless the FDR is tampered with, the data should contain the last flight.

4. Data analysis of the FDR only gives the following information:

   a) The total duration of data recorded is for 59 minutes.
   b) The date, time and the airbase airborne or landed cannot be ascertained.
   c) The record at 55th minute indicates the speed of 82 km and direction 264.00 at a height of 453.36 meters. Hence it may be presumed that aircraft had landed at an airfield having runway direction of 264.00 (approximately) and situated approximately about 453.36 meter [sic] above sea level.[41]

The CBI's chargesheet alleged that 'investigation [had] established that Ananda Marga monks [Tadbhavananda, Saileshwarananda, Jagdishwarananda and Suranjanananda] entered into a conspiracy to wage war against the state to implement a new-humanistic world government in terms of the Proutist philosophy propounded by the founder of the Ananda Marga, Prabhat Ranjan Sarkar [also known as Anandamurti]'. According to the chargesheet, 'it was revealed that on September 29, 1995, at meeting of the Central Committee of the Ananda Marga Pracharak Sangha in Calcutta, Jagdishwarananda and Tadbhavananda announced certain proposals which, inter alia, include(d) "rule the country to get economic power etc".'[42] Besides, the investigating agency claimed that Tadbhavananda, who was then secretary general of Proutist Universal, had mentioned in one of his reports that 'the advent of the fourth phase will be preceded by a third phase in which there would be much human loss, suffering and turmoil. But we will have to wage war against the immoral forces'. During the meeting, it was announced that the Ananda Marga 'would participate in the 1996 parliamentary elections for which they had formed an alliance with two local [political] parties with a decision to defeat the commies [the communists who were in power in West Bengal] and to get rid of them'. The chargesheet alleged that some of the seized documents 'speak about arms training imparted during August 1995 at Ananda Nangar and further plans of continued training in December, 1995'.[43]

The so-called evidence—'neo-humanistic world government', 'ruling the country to get economic power', participating in the 1996 parliamentary elections and arms training—and the attempt to link them to the arms drop was, to say the least, bizarre. In its wisdom, the CBI believed that a section of the Ananda Marga monks sought to achieve these aims by smuggling in large quantities of arms and ammunition, which they would use to wage war against the state to attain the objectives allegedly propounded by Tadbhavananda and some of his monkish colleagues. The CBI next sought to link the arms drop to the discovery of a book titled *Techniques of Modern War* and an article that had appeared in the *Indian Defence Review* magazine, which were seized during search operations at the Purulia headquarters of the Ananda Marg. It claimed in its chargesheet that these discoveries 'reinforce the evidence against the accused persons to establish their intention to wage war'[44].

There were other hare-brained attempts to link the Ananda Marga monks to the arms drop. For instance, from a bunch of emails that the CBI could extract from the computers seized during its search of the Ananda Nagar premises, the agency picked one containing the word 'weapons'. The email, purportedly written by one Annabel Perkins alias Jayashrii (who lived in the US and was associated with Cornell University), to another senior monk (suspected to be Tadbhavananda), questioned how these weapons would be tested and when would they be used. The use of the term 'weapons' was conveniently, but mindlessly, linked to the expression 'wage war' and was considered to be sufficient evidence to chargesheet the four monks.

There was no doubt that the four avadhoots were linked to the arms drop. Nothing else explains their disappearance from Ananda Nagar after the arms drop and their continued run from the law.[45] But the evidence the CBI gathered against some of them, including Tadbhavananda, Saileshwarananda and Jagdishwarananda, was a desperate attempt to make the case stick against them. If there was other sensitive evidence, the CBI thought it prudent not to highlight them in the chargesheet. Besides, it took the plea that the arrest of the Ananda Marga monks would shed more light in support of its case. A supplementary chargesheet filed by the CBI after Tadbhvananda's arrest from Delhi on 15 April 2004 indicated that after the arms drop, the Proutist Universal secretary general left the Ananda Nagar ashram, travelling to various places and staying in Ananda Marga institutions, changing his name. 'In Aurangabad and Dehri-On-Son [both in Bihar], he called himself Acharya Ramkrishnananda Avadhoot,' the chargesheet, filed in court on 16 November 2004, said.[46]

According to the supplementary chargesheet, the MEA (File No. VI/405/1/255/77 of 11 November 1977) had placed Tadbhavananda in the 'prior approval category' because he was suspected to have committed 'dangerous and unlawful activities' that could potentially destabilize the country. The MEA order, therefore, effectively made him ineligible for any Indian passport in his name or under his other aliases such as Tushar Kanti Parihar or Lal Chand Parihar, without the prior permission of the ministry. But the CBI found that while Tadbhavananda did not avail of any passport, documents seized from a few Ananda Marga institutions made it clear indicate 'that in pursuance of Prout[ist] work, he had travelled to Bangkok in 1992, Copenhagen in 1993, Taiwan and a few other countries, indicating he used forged/false travel documents to make these international visits'[47]. During interrogation while in CBI custody, Tadbhavananda denied knowing Nielsen or any of the other absconding Ananda Marga monks, barring Jagdishwarananda. He disclosed that from January 1996, he was in hiding in Delhi, Varanasi, Azamgarh, Gorakhpur, Nepal, Gujarat, Kanyakumari, Belgaum, Goa, Mumbai, Aurangabad, Bokaro, Ranchi and Jamshedpur.

A suggestion was made that on the night of 17 December 1995, Vinay Singh, Suranjanananda, Saileshwarananda and Jagdishwarananda, among others, were tasked by Nielsen to flash a powerful light from the terrace of the two-storeyed white building in Bansgarh to the AN-26 aircraft at the time it was to airdrop the weapons. The pilot, it was claimed by the prosecution, missed the signal, mistaking the lights further south, where work on the railway tracks at Kotshila was going on, as the actual target. Recall that the third pallet thrown out of the AN-26, which flew in as low as 1,500 feet from the ground (and thereby avoided being spotted by the nearest IAF radar station at Kalaikunda) after it deviated from its course over Gaya in Bihar, landed way off target. Nielsen's use of the GPS probably contributed to the parachutes overshooting the intended target area. The drop began when the GPS indicated that aircraft was directly over the intended target, although it should have commenced much earlier to allow for drift and the aircraft's forward motion.

Besides, on the afternoon of 17 December 1995, an Ananda Marga truck had met with an accident, running over a couple of children at Khatanga village. Pranab Kumar Mitra recalled, 'As a result, agitated villagers tried to beat up the driver who managed to flee the accident site. When the incident was reported at Jhalda police station, a Sub-Inspector, Dipankar Bakshi, was sent to the spot. Bakshi, who reached Khatanga well after sunset, engaged a few villagers to guard the vehicle while he returned to the police station to prepare the papers related to the accident.'

According to Mitra, the presence of villagers in Khatanga, which is barely five and a half kilometres away from the white building, 'appeared to have prevented the persons waiting by the white building to collect the arms and ammunition once the parachutes drifted down on the open fields late that night.'[48]

The trial court was circumspect about the involvement of the Ananda Marga organization as a whole. In its judgement, delivered on 31 January 2000, the court pointed out that 'no sufficient evidence could be adduced' by the prosecution to link the Ananda Marga organization as a whole 'with regard to the present matter' and that 'it will be too much [for the court] to hold that all the members of Ananda Marga were involved in the commission of the offence'. However, it noted that

> ...it should be taken to have been well established that the arms and ammunition were brought to be delivered through air-dopping to some persons of Ananda Marga Pracharak Sangha at a targeted place at Bansgarh...may be persons linked

up with the present process may have nexus with the Ananda
Marga organisation...I hold that it has been established from
the materials on record that the places where the arms were
targeted to be dropped were of (the) Ananda Margis.[49]

The court asserted that the prosecution could not 'adduce any
convincing evidence with regard to the involvement of accused Vinay
Kumar Singh with regard to the present matter.' It agreed with the
prosecution that Vinay Singh was associated with Ananda Marga, but
noted that 'there is no cogent evidence and materials on record to
prove [the] involvement of Vinay Kumar Singh with the concerned
crime.'[50]

Vinay Singh was acquitted of all charges.

In the initial days, the trial, which lasted slightly less than three
years, would attract throngs of people and advocates eager to catch a
glimpse of the Latvians and the nattily dressed Bleach. On every trial
day, Bleach would be brought to the Bankshal Sessions Court in the
heart of Calcutta in a ramshackle Calcutta Police van from Presidency
Jail, where he and the Latvians were incarcerated as undertrials after
their arrest in Mumbai.

Bleach did not engage any advocate to represent him, contesting
the case on his own. During the weeks and months of incarceration at
Presidency Jail, he familiarized himself with Indian criminal law,
various legislations that governed the Criminal Procedure Code and
the Law of Evidence. He would draft his brief, besides writing letters
to benefactors in the UK, including a few Conservative party MPs
and other British officials—all in a bid to impress upon the Indian
court that he was innocent, had kept the British police informed
about the conspiracy to procure the weapons and had fully cooperated
with the CBI during his interrogation and in the course of the
investigation. Only during moments of frustration he would vent his
anger against the CBI, as reflected in a letter he wrote to then British
Deputy High Commission second secretary in Calcutta, David
Belgrove.[51] In the eight-page letter, Bleach requested Belgrove to
testify in court so that his life was saved. He wrote:

As you are aware, some of the circumstances surrounding my
involvement in this affair are in dispute. It is, however, not in

dispute that I informed the appropriate government authorities about this situation immediately that I myself became aware of it. There was therefore a very clear British government involvement of some kind from the very outset...

[...] The CBI has now informed the court, in writing, that I did not assist them at all after my arrest, and that I did not make any statement. Of course I maintain that I supplied a statement of around 40 pages, detailing every aspect of the affair, typed by me personally on Superintendent Behera's computer...

[...] In fact 'Davy' remained in India for at least a week after my arrest, and it is my belief that he only departed after seeing the extent of my cooperation with the CBI reported in the press.[52]

Bleach was indeed right. He did cooperate with the CBI, almost fully. Immediately after his arrest and during several rounds of interrogation—separately—by not just officers of the CBI, but also the R&AW and the IB, he shared copious details related to the purchase of the weapons and the aircraft. With the benefit of hindsight and in the light of documentary evidence that emerged subsequently, the CBI misrepresented in court that the Briton had not cooperated. On the contrary, Bleach's insights into the case—the conspiracy as well as the execution of the arms drop operation—considerably helped the CBI proceed with its investigations and prepare a prosecution argument that, unfortunately, nailed Bleach as well as the Latvian crew members, who knew nothing at all of the conspiracy, though they did know that weapons were being carried on the AN-26 for clandestine delivery in India. When the R&AW's Mumbai-based special commissioner, Jayant Umranikar, interrogated the crew in Russian, it emerged that they had expressed extreme displeasure over the nature of the cargo after the AN-26 took off from Karachi, but went ahead with the flight in obligation of the contract they had signed with Nielsen. 'They did it for the money and the promise of a great vacation in Phuket,' Umranikar told me during a telephone conversation, adding, 'All of them were very experienced aviators and the pilot, Alexander Klichine, had once flown the former Soviet President Mikhail Gorbachev.'[53]

Financial Aspects of the Deal

A critical and vital part of the investigation—and perhaps the most painstaking and arduous—was tracking the movement of money and the financial dealings between Nielsen, Latavio, Latcharter, Border Technology and Innovations (BTI) and the Bulgarian arms manufacturing companies. The long and tedious process to get to the root of the transactions also had the spin-off benefit of establishing the identities of not just the companies involved but the individuals who actually spent the huge amount, as well as the recipients. To simplify the complex web, the financial aspects were classified under five broad heads: procurement, transportation, hiring of the crew, purchase of accessories and other miscellaneous expenses.

While I have dealt with some of the transactions in Chapter 4, information gathered in Bulgaria showed a transaction amounting to $148,000 which was paid by way of four demand drafts drawn on a bank in Switzerland. While $25,000 was paid by a draft (No. 6229595 of 22 December 1995) drawn on the Bank of Switzerland in the name of KAS Engineering, Sofia, a sum of $48,000 was paid to KAS Engineering by way of a draft (No. 6229696 of 22 December 1995) again drawn on the Bank of Switzerland. Two other drafts (No. PPS 493793 of 10 November 1995, and PPS 493972 also dated 10 November 1995), totalling $75,000 and drawn on the Union Bank, Switzerland (Account No. 8462A9-600-BC-278), were paid to BTI. A separate payment of $10,000 was probably made to Peter Bleach for his mediation in the deal, but the CBI had no clue to indicate the instrument of payment to Bleach. Other documents the CBI could procure from Bulgaria suggested that $85,000 was paid in advance and an identical amount paid by draft, i.e. a total of $170,000 was paid for the purchase of the arms and ammunition. This apart, a total of $11,085 was spent as wages towards the five Latvian crew members. BTI's profit as the broker between Nielsen and KAS Engineering Co. stayed at 116 per cent.[54]

As for the purchase of the AN-26, initially a contract was signed on 1 November 1995, between Latcharter UK Ltd and Carol Air Services Ltd for $250,000. The payment condition was either by cash or by bank transfers. The contract papers reflected the beneficiary's (Latcharter Airlines UK Ltd) bank as Midland Bank PLC, 6, Broad

Street, Worcester, WR1, account No. 37306843. Subsequently, a purchase contract was signed between Kim P. Davy and Mara Bekere who represented Latavio Latvian Airlines, on 15 November 1995, when it was decided to purchase the AN-26 at a cost of $180,000. The contract mentioned payment by bank transfer to an account at the Swiss Bank Corporation, 6, Parade Platz, Switzerland, Swift Code SBCOCHZZ 80 A, and the beneficiary bank was Parekss Bank Corporation with Account No. PO-1678442, in favour of JLY Enterprises Corporation, which was supposedly the agent of Latavio.

Investigations showed that Latavio received an amount of $160,000 in its account with Bank of Baltija by draft drawn on the Bank of Parekss. It was suspected at the time that there werepossible ways the amount was paid: by a draft that was purchased from the Bank of Parekss by paying cash after which the draft was deposited at the Bank of Baltija; or in accordance with the purchase condition, the amount was transferred from the Swiss Bank Corporation to the account of JLY Enterprises Corporation's Bank of Parekss. JLY Enterprises in turn arranged to get a draft after deducting their commission and the final amount of $160,000 was deposited in the Bank of Baltija account of Latavio Latvian Airlines.

In the course of investigations, the CBI came across a code CITCO (SUISSE), which was believed to be that of a bank in Switzerland, along with fax number (41223101948). Although the Swiss authorities refused to divulge banking information to the CBI, the investigating agency strongly believed that CITCO was part of a code to an Alpha account in a Swiss bank and held the key to not just the transfer of the amount for the purchase of the arms and ammunition, but also the name of the account holder. According to Bleach's statement, Nielsen had made a telephone call to Hong Kong, following which the amount was transferred from a Hong Kong bank to a Swiss bank which subsequently paid KAS Engineering. The CBI's investigations took the agency to a lead that sometime after the arms drop and his escape from Mumbai airport, Nielsen had contacted a finance broker, one Mrs Singh, in Palm Springs, California—an indication that he had financial links in the US.

The Letters Rogatory sent by the Calcutta trial court to 'competent authorities' in the UK, Bulgaria, Latvia, South Africa, Hong Kong,

Singapore, Thailand, Taiwan, Bangladesh and Denmark, to examine, interrogate and elicit information from individuals who were aware of the purchase of the weapons and the aircraft, threw substantial light on the arms deal operation and the manner in which it was executed. However, the Indian request to Pakistan for assistance in the investigation met with cold reception. By 2001, the CBI's investigation had begun to crumble at its seams, especially after Interpol's Project Purulia assignment to determine the source of the weapons and the details of their purchase and to pinpoint the whereabouts of Niels Christian Nielsen was over. There were no further leads on either the fugitive Ananda Marga monks or on Nielsen, especially those involved with him at Howerstoke Trading Ltd, the Hong Kong-based front company which was suspected to be at the forefront of the Purulia conspiracy, and was listed as the buyer of the airdropped weapons. More alarmingly, the CBI could not trace some of the missing weapons and equipment.

Arms/Ammunition	Dropped	Recovered	Missing[55]
RPG-7 rocket launchers	10	10	—
AK-47 M1	300	247	53
Makarov 9-mm pistols	25	11	14
Dragunov 7.62 sniper rifles	2	2	—
Night vision binoculors	2	2	—
Grenades offensive	100	78	22
7.62 ammunition	23,800	20,543	3,257
9-mm ammunition	6,000	3,885	2,115
Anti-tank grenades	100	78	22
Telescopic sights for rocket launchers	10	6	4

From Life Imprisonment to Freedom

When the trial ended on 1 January 2000, the judgement delivered by the sessions judge held Peter Bleach and the five Latvians guilty of the offence of conspiracy to wage war against the state. They were also charged under provisions of the Arms Act, 1959, the Explosives Act, 1884 and the Aircraft Act, 1934. They were, however, not found guilty of having committed any offence of waging war against the state. A month later, when it was time for the court to pronounce sentence on the six convicts, it observed that the offences committed by them 'cannot be sheltered with any kind of mercy' and sentenced the six convicts 'to suffer rigorous imprisonment for life' along with fines ranging from Rs 3,000 to Rs 25,000 each.

Long before the conviction and reading out of the sentence in the Calcutta court, civil rights groups in Russia had begun applying pressure on the Vladimir Putin government to seek the release of the five Latvians from Presidency Jail. Some of the Latvians had lost weight and suffered various diseases in the unhealthy and unhygienic environs of the jail. Bleach, too, was struck by tuberculosis and had to be hospitalized. The Latvians were allowed to meet their wives when they arrived in Calcutta, soon after the news of their arrest was splashed on the front pages of every newspaper across the world. Most of the wives spoke only Russian and had to communicate in gestures and sign language to proclaim the innocence of their spouses. They virtually ran from pillar to post, pleading with Indian officials, the political leadership and the media, that their husbands were decent, law-abiding citizens of Latvia; that they loved India and wanted no harm to come to it.

The National Democratic Alliance government led by the BJP tried to resist pressures from the Russian government, but subsequently buckled when Putin threatened to cancel a state visit to India if New Delhi did not grant presidential clemency to the Latvians. The Russians argued—and rightly so—that the Latvians had been tricked into the deal and had joined the employment of Nielsen's Carol Air Services Ltd purely as a means to better livelihood opportunities. Although the Latvians did learn later that the aircraft was carrying lethal weapons for delivery to India, they could do little to pull themselves out. They went along with it, partly fearing for their lives and partly to make the most of a flying adventure. On 22 July 2000, three months before Vladimir Putin's visit, as a goodwill gesture, the Latvians were granted presidential pardon by an executive order of the Indian government.[56]

As in the case of the Latvians, a powerful campaign had built up in England, led by the then Conservative Member of Parliament for Rochford and Southend, East, Sir Teddy Taylor, seeking Bleach's release from Indian prison. Taylor's support for Bleach's release began soon after stories about the arms drop appeared in the British media. Taylor relentlessly kept up the pressure on the Labour government of Prime Minister Tony Blair, raising the matter in the House of Commons and other fora to pursue what he described as a 'great injustice'[57]. On several occasions Taylor said he was 'seriously concerned and alarmed about the health and security problems which Peter is having in a prison in Calcutta' and his objective 'in raising the matter in this debate is to try to persuade the Minister (the then Home Secretary Jack Straw) and his Department to do all they can to secure Mr Bleach's release.'[58]

So strong was his concern for his friend Bleach that in 1997 he was reported to have almost agreed to a plan to hire a British mercenary to free the Briton from Presidency Jail. *The Mirror* reported that on 21 September 1997, Taylor 'plunged into an amazing row over a shadowy plot to free a British arms dealer from an Indian jail. He was accused of trying to recruit a mercenary, John Miller, a former Scots Guard, to spring Peter Bleach.'[59] Taylor did not deny the plot, saying of Miller that 'we met in my office where he revealed a complex plan to free Peter, which was so far-fetched I did not take it very seriously.'[60] He subsequently took to more democratic and legitimate means to seek Bleach's release from Presidency Jail.

'Peter (Bleach) was no angel in his past business dealings' and in the Purulia episode he 'was guilty of many things. He was a fool to have got himself entangled in the mess,' Christopher Hudson, an old friend and political associate of the British arms dealer from North Yorkshire, told me over a long telephone conversation.[61] In the summer of 1997, Hudson moved the Foreign Office and secretly met with the then MI6 director of operations, Richard Dearlove (promoted in 1999 to director general, heading the agency, and subsequently knighted), who 'gave a categorical assurance that Bleach's life would be saved.' Dearlove, however, denied—not surprisingly—having 'any knowledge whatsoever of the matter (the arms drop)'.[62] Hudson also spoke 'for ten minutes' to a Labour member on the Intelligence and Security Committee, Dale Campbell-Savours (later to become Baron Campbell-Savours) who assured him 'that the information would be passed on to the appropriate authorities'.[63]

By 2002, the Blair government had gone out of its way to try to persuade the Indian government that injustice had been done to Bleach. Blair personally spoke to then Indian Prime Minister Atal Behari Vajpayee seeking the BJP leader's intervention. Then British Foreign Secretary Robin Cook raised the issue with Indian Deputy Prime Minister and Home Minister L.K. Advani on 22 August 2002. Another Foreign Office minister, Under-Secretary of State for Foreign and Commonwealth Affairs and MP for North Warwickshire Michael O'Brien had discussed the matter with Indian Foreign Secretary Kanwal Sibal on 16 October 2002.

The release of the Latvians provided Bleach the grounds to seek clemency too. From his prison cell, he drafted letters and petitions to the British government and the Indian judiciary, demanding that he be treated at par with the Latvians. A court observed in *Peter Bleach Vs State of West Bengal and Others*:

> Peter James Gifran von Kalkstein Bleach, a British national, appearing in person, petitioned before this (Calcutta) High Court with the grievance that he is a victim of discrimination as the President of India, while acting in exercise of power under Article 72 (of the Constitution), refused to remit the sentence imposed on him while granting the said prayer of five other Russians who were similarly convicted and sentenced with him.[64]

The court of Justice A.K. Ganguly rejected Bleach's petition on 22 September 2002, while holding that the trial court's judgement

> ...established beyond any reasonable shadow of doubt that accused Peter Bleach played a vital role in purchasing the

aircraft, engaging the Latvian crew members and using the contract number of Bangladesh (End-User Certificate) and arms were loaded in the aircraft mentioning those as 'technical equipment' and those were thereafter smuggled into Indian territory with (the) active connivance of the Latvian accused persons.[65]

The Indian government, more precisely the MHA, sought to adopt a stand tougher than that of the MEA under Jaswant Singh. The MHA declined to respond to the requests of the Blair government, saying that 'the case for remission/premature release of Peter Bleach had been examined and considered in this Ministry in consultation with all concerned, including the CBI, and it was then decided not to release him in view of the role played by him.'[66] The MHA argued that

the role of Peter Bleach in the conspiracy is different in its gravity than that of Latvians. Bleach is the co-founder of the conspiracy [and] was instrumental in obtaining the aircraft and the arms and ammunition along with another key absconding accused, Kim P. Davy... He was also instrumental in obtaining fraudulently an End-User Certificate from the Bangladesh Government and supplying the same to the Bulgarian authorities to create an impression that he was procuring the arms for the Bangladesh Government.[67]

The MHA concluded that 'Peter Bleach does not deserve consideration for clemency by the President under Article 72 of the Constitution of India. There are no mitigating circumstances too in favour of the accused person.'[68]

However, the Indian government could not withstand the British insistence and pressure to secure Bleach's release. It appeared that the political leadership went along with External Affairs Minister Jaswant Singh, whose view prevailed over that of his ministerial colleague at the MHA, Deputy Prime Minister L.K. Advani. Yet again, like its officials who failed to take timely action on the R&AW's intelligence, displaying themselves as incapable of taking tough decisions, the political leadership exposed itself as soft and jelly-like, succumbing to British pressure and making a mockery of the Indian judicial system, whatever it was worth.

When Bleach walked out of Presidency Jail a free man on 5 September 2004, four days after British Home Secretary David Blunkett's meeting with Advani, he was attired in a navy blue blazer and Ray Ban sunglasses, and had grown a luxuriant moustache. Back in England, though, his business disintegrated. Without friends (his girlfriend Joanne Fletcher, too, had grown distant) and shunned by his business colleagues, Bleach was alone, feeling betrayed and double-crossed by British intelligence and the Indian authorities.

9

THOU SHALT NOT GET CAUGHT

Eighteen days after escaping from Mumbai airport in the wee hours of 22 December 1995, Nielsen crossed over into Nepal and out of the grasp of any Indian law enforcement or intelligence agency. At that time, the CBI had no clue about Nielsen's whereabouts or his subsequent escape route. CBI investigators would learn about Nielsen's crossover to Nepal months later, with assistance from R&AW, whose local operatives in Mumbai had discovered that after emerging from the airport terminal building, Nielsen had placed three telephone calls to trusted contacts in Hong Kong and Pune. He then walked out of the airport terminal and roused a Fiat-Premier Padmini cab driver, Arvind Dharamdas Pal, from his sleep, directing him to head for Ambassador Hotel in Churchgate, where he reached around 0545 hours. Nielsen did not enter the hotel, instead anxiously loitering on the pavement, apparently waiting for a rendezvous with someone. When his contact, either an American or European, reached the hotel, the two hired a taxi and headed for downtown Mumbai. Nielsen stopped the taxi at Al Amin Communication Centre on Nowroji Furdunji Road, using their pay-phone to make telephone calls to his landline phone in Hong Kong. Within ten minutes of seeking a callback facility, Nielsen received a call from a Hong Kong number. The pay-phone operator noted that Nielsen was accompanied by another 'white-skinned' man. Nielsen also used the pay-phone services at Waghela Communication, a few blocks away on the same road, to make more phone calls to Hong Kong. The two men then hired another cab and headed for the Osho Commune International ashram in Koregaon Park, Pune.

On reaching Pune, Nielsen and his associate checked into the Osho Commune International guest house without any difficulty or questions asked.[1] After spending over a week at the Osho ashram, where newspaper and television news reports provided him with information on how much knowledge law enforcement agencies had of his whereabouts, Nielsen hired another car and headed for Maharajganj (a district in Uttar Pradesh) on the India–Nepal border, a distance of 1,403 kilometres. Since his photographs were yet to be splashed over the front pages of newspapers, taking the land route instead of flying or boarding a train was the safest bet for him to travel the distance to the India–Nepal border unnoticed.

Indian intelligence has, time and again, been found to work best post-event. It was no different when the manhunt for the phantom Dane began, which involved Interpol, police and intelligence agencies across five continents. A couple of months after Nielsen was suspected to have crossed over into Nepal, the IB shared information with the R&AW that an American, Robert Michael Norton (US passport No. Z-6470713, issued on 21 June 1988 at Jakarta, Indonesia), along with one Michael Shane Tapp, had checked in at Niranjan Hotel, Maharajganj, on 9 January 1996. They had appeared at the Indian immigration check-post at Sonauli in Maharajganj district the next morning for proceeding to Nepal. However, both of them suddenly left the check-post, leaving behind the travel documents of Michael Shane Tapp, on the plea that they immediately needed to consult a local doctor. The duo never returned to the immigration check-post to collect Tapp's travel documents.

The Manhunt Begins

At this point, R&AW stations in Nepal and a few other countries, including Venezuela, were alerted by the agency's New Delhi headquarters to pursue the matter and follow up on the lead that the IB had provided. The R&AW's inquiries revealed that a Michael Shane Tapp had lodged a complaint at Paharganj police station in New Delhi on 9 January 1996 at 1430 hours, saying that he had lost his New Zealand passport (No. L-135381) two days ago, somewhere in Paharganj's crowded main market. At the same time, the arrival into India of the American national, Robert Michael Norton, who had accompanied Tapp to the Sonauli check-post, could not be found in immigration records.

It was later confirmed by the R&AW that the New Zealand passport, no. L-135381 in the name of Michael Shane Tapp, was false. It was obtained in the same manner as the New Zealand passport in Kim P. Davy's name, i.e. by using the name of a dead infant. It was also subsequently established that Tapp was identical with New Zealand citizen David George Hammond who had disappeared from New Zealand in August 1993. It was Hammond's fingerprints that were lifted from Tapp's Colombian identity card, which was used as proof of identity when the individual claiming to be Tapp applied for a new passport at the New Zealand High Commission in New Delhi on 10 January 1996. As the R&AW dug further, more and more information came to be revealed, this time in Columbia. Inquiries showed that 'Tapp' was a resident of Bogota, the Colombian capital, and was booked to fly to São Paulo on 19 January 1996, a booking which was subsequently cancelled.

Meanwhile, Robert Michael Norton, who had disappeared from the Sonauli immigration check-post along with Tapp on 10 January 1996, was identified as Steven Michael Dwyer. He was involved in the stabbing of an Indian Embassy official in Manila on 7 February 1978.[2] He was also known by the Ananda Marga spiritual name of Meenakshi Sundaram, alias Arghyananda, alias Arisudas. Born on 24 March 1949, Dwyer initially obtained an American passport (no. C-0447602) on 23 March 1972 in his own name. Later, he obtained passport no. Z-2723887 in Kuala Lumpur on 14 December 1977. He acquired yet another passport (no. Z-2682597) issued on 8 December 1977, under his original name. The next passport obtained by him was no. Z-4125700 issued in Manila on 21 February 1982. He got his name changed to Robert Michael Norton in Denver, Colorado, on 28 February 1982. On the basis of this change in identity, he got his name on passport no. Z-4125700 altered to Norton in Bangkok on 13 July 1984. He procured yet another passport (no. Z-5877542) issued at Bangkok in the name of Norton on 13 November 1984. A sixth passport (no. Z-647713) in the name of Norton was obtained from

Jakarta on 21 April 1988, and it was this passport which was presented at the Sonauli check-post.

Sometime in October 1996, the Indian mission in Caracas, Venezuela, received an anonymous telephone call informing them that Robert Dean Child, Nand Kumar and others connected with the Purulia arms drop were on their way to India. The São Paulo telephone number of Nand Kumar (not an Indian national), who was the sector secretary of the Ananda Marga for the Georgetown sector and was based in São Paulo since 1986, was shared by the anonymous caller. The R&AW's inquiries now revealed that Robert Dean Child, at the beginning of 1996, used the same telephone number (+55-11-2047954). It was also learnt that Robert Dean Child was a new name adopted by Robert Dean Coddington Jr.

Child had earlier used the names of Victor Daniel King, Priyadarshi, Proshidananda Dada and John Paul Christians. After assuming the name of Victor Daniel King, Child was in Southeast Asia for fourteen years at the Ananda Marga's headquarters in Manila. He had also come to the notice of Indian intelligence for making threatening statements against the then prime minister, Indira Gandhi, and the Indian government sometime in 1973. During the period when Child was based in Manila, the fatal stabbing of an Indian Embassy official on 7 February 1978 occurred. At that time, Steven Michael Dwyer, the Ananda Margi who had stabbed the Indian official, had the name Robert Michael Norton.

The R&AW was able to ascertain that Robert Dean Child entered India around 24 December 1995, left via Indira Gandhi International Airport, New Delhi, on 3 January 1996, re-entered India two days later and left via Sonauli on 9 January 1996.

It was indeed a strange coincidence that two individuals, Norton and Robert Dean Child, both known Ananda Margis with a history of violence, both based in Manila at the same time and both apparently known to Nielsen, passed through or attempted to pass through the Sonauli immigration check-post on 1 January 1996. For Indian intelligence, yet another coincidence was that 'Michael Shane Tapp' got his false New Zealand passport in the very same manner as Kim Palgrave/Peter Davy (Nielsen) got his false New Zealand passport. It was also strange that Tapp was booked to travel to São Paulo, where Child was based at that time. Another coincidence was that a piece of paper bearing the telephone number 55-11-2047954 (Child's São Paulo number) and the words GTSO/John was found in Nielsen's Hong Kong apartment after the Purulia arms drop[3]. GTSO could very well be an acronym for Georgetown Sectoral Office, and John was one of the names which Child had earlier used (John Paul Christians). It was another coincidence that Nielsen's business associate Angel Agonal Caparaz, a Filipino from Manila who used to operate from Nielsen's Hong Kong apartment, was known by the name 'King', the same name which Child used when he was based in Manila.

These coincidences strongly suggested two possibilities, according to the R&AW. One was that the person who presented Tapp's passport at the Sonauli check-post and then left in a hurry was Kim P. Davy. For the R&AW, the following points led to the suspicion:

1) Michael Shane Tapp's original Colombian identity card was presented at the New Zealand High Commission in New Delhi on 10 January 1996, when an application was made for a new passport. Since the identity card contained his fingerprints, it was presumed that the person who appeared at the New Zealand High Commission was the real Michael Shane Tapp.

2) It followed that the person who presented the passport of Tapp at the Sonauli check-post on 10 January 1996 was not Tapp.

3) However, that person was connected with the Ananda Marga because he was accompanied by Norton, a well-known Ananda Margi.

4) There was no need for Michael Shane Tapp, who had entered India on a valid travel document, to seek a new passport in his own name unless he had genuinely lost his passport or had willingly handed it over to a friend of his to help him leave India.

5) The person who presented Tapp's supposedly lost passport at the Sonauli check-post, being an Ananda Margi, was highly unlikely to have stolen it from Tapp in the Ananda Marga. Therefore, in all likelihood, it was handed to him willingly by Tapp because he (the person who presented Tapp's passport at Sonauli check-post) could not leave India on his own passport.

6) The only foreigner connected with the Ananda Marga who could not leave India on his passport at that point in time was Nielsen. As far as R&AW operatives were aware, there was no specific advice/information about any other foreign Ananda Margi who was in India at that time, which would have warranted his apprehension at the border.

However, one factor which did not fit into this theory was the fact that from the copy of Tapp's passport photograph available with the R&AW, it appeared that Tapp had dark hair, whereas Davy had blond hair. It was learnt that the US Diplomatic Security Service was able to establish a positive identity of Kim P. Davy as Niels Christian Nielsen on the basis of fingerprints lifted from a US Customs form on 18 July 1995 in Newark, New Jersey, which matched those of Nielsen provided by the Danish authorities to their Indian counterparts. At that time, it appeared to Indian intelligence that the technology to lift fingerprints from plain paper existed. The R&AW, therefore, advised that it would be worthwhile to check whether any fingerprints matching those of Kim P. Davy were seen on the passport of Michael Shane Tapp which, the intelligence agency believed, was available with the Uttar Pradesh police.

The second possibility was that the person who claimed to be Norton at the Sonauli check-post was Kim P. Davy. The circumstances which lent credence to this theory were:

1) Davy had made repeated telephone calls to his Hong Kong apartment after his escape from Mumbai airport on 22 and 23 December, 1995.

2) In all likelihood, it was 'King', Davy's associate, who took these phone calls at the Hong Kong flat. King was from Manila, where Robert Dean Child had worked for fourteen years using the name 'King'.

3) A piece of paper with Robert Dean Child's São Paulo phone number, one of his aliases (John) and the name of the office to which he was apparently attached, was found in Davy's Hong Kong apartment after the Purulia arms drop.

4) It was, therefore, likely that King received the message that

Davy was in trouble in India, contacted Child in São Paulo and requested him to proceed at once to India.

5) Child was known to have travelled to Mumbai on 24 December 1995.

6) The arrival of Robert Michael Norton, who disappeared from the Sonauli check-post on 10 January 1996, was not found in immigration records. It was, therefore, strongly believed by R&AW that Norton's passport was brought along by Child to help Davy escape from India. The fact that the person who claimed to be Norton appeared at the Sonauli check-post on 10 January 1996, a day after Child exited through the same check-post, reinforced this suspicion.

In the backdrop of these startling findings, the R&AW suggested to the IB and the CBI to pursue the following line of investigation:

1) Arrangements should be made for examining Michael Shane Tapp's passport for fingerprints of Nielsen.

2) It was learnt that Child had again changed his name. While the new name was not known at the time, the passport number was believed to be Z-6470713. It was considered unlikely that he would have visited India with this new passport. A search of the immigration records for US passport no. Z-6470713 might reveal his new name.

3) Similarly, it was suggested that details of Child's onward journey from Nepal would also throw more light on his suspected role in assisting Nielsen to escape from India. It was found significant that Child thought it prudent to change his name to Robert Caring on 22 December 1996 (i.e. in less than two weeks after he crossed over to Nepal through the Sonauli immigration check-post on 1 January 1996) and to have obtained a fresh one-year multi-entry visa for India on 23 January 1996. Using that visa, he visited India nine times during the course of 1996.

Barring a positive identification of Nielsen's fingerprints on Tapp's passport, which was available with the Uttar Pradesh police, the CBI could verify all of the information that R&AW had supplied on the fugitive's secretive border crossing. Though fraught with danger, Nielsen's border crossing was a remarkable operation, an ingenious mix of artful deception, skilful use of available resources, meticulous planning and calculated risk-taking. He seemed to have followed every intelligence operative's Eleventh Commandment: thou shalt not get caught. Child and Nielsen took a flight from Kathmandu's Tribhuvan International Airport on 19 January 1995, reaching Frankurt the same day. At Frankfurt, Nielsen took a flight to Copenhagen and Child to the US, most likely Colorado. Nielsen employed good tradecraft, which made it difficult to locate him.

A CBI officer involved in the investigation said that contrary to Nielsen's claim on Times Now, an Indian television news channel, on 28 April 2011, the fugitive Dane did not take the help of Bihar MP Pappu Yadav to escape to Nepal. Yadav, as also his father, was close to the Ananda Marga and had in the past attended at least one Dharma Maha Sammelan, but he played no part in transporting Nielsen to Nepal. Nielsen took Yadav's name since it had been mentioned to him by Randy, who too had close links with the cult organization. It was all part of an elaborate ruse to sensationalize the issue and deflect attention from him. At the same time, Nielsen's allegation, made in the interview with Times Now, that Yadav was instrumental in using his political clout to have the IAF radar at Kalaikunda shut down on the night of 17 December 1995, was another means to project the alleged nexus between politicians and officials who, he claimed, had planned to bring in weapons to India.

As 1996 wore on and nothing much could be known about Nielsen's whereabouts, the CBI enlisted the support of the Interpol, with which it had a very close working relationship. Interpol's Analytical Criminal Intelligence Unit (ACIU) chief, Christer Brannerud, met CBI SP Loknath Behera in Calcutta and the two teamed up to pursue the manhunt. Brannerud followed an ingenious method, though with some difficulty, to track down Nielsen's movements, by scrutinizing his expenditure pattern as it emerged from the use of his Citibank Visa card.

A day after escaping from Mumbai airport, among the telephone calls that Nielsen made was one to Hotel Pearl Intercontinental in

Karachi, where the two ground engineers, Vladimir Ivanov and Alexander Lukin, had been advised to stay put till the AN-26 returned to the Pakistani city. He told them to return to Riga. On 1 January 1996, a payment of $357 was made to Cathay Pacific Airways, indicating purchase of an air ticket. On 19 January 1996, by which time he had escaped from India after being suspected to have crossed over to Nepal, a transaction of $772 was noticed to have been made to Alley Photo Traders in Singapore.[4] These were the two last transactions recorded on Nielsen's Visa account. Tracing the transactions backwards revealed to the ACIU the dates and places where Nielsen travelled to before the 17 December 1995 arms drop operation was executed. This proved to be invaluable for the CBI, which had little expertise at the time to track the money 'footprints' of Nielsen.

Between 1 January and 5 December 1995, Nielsen flew in and out of Hong Kong on at least sixty occasions. Going backwards, the ACIU discovered that the first of Nielsen's many Visa card transactions was in Seoul, South Korea, where on 14 December 1994, a payment was made to Asiana Airlines. Three days later another card payment was made to Cathay Pacific in Hong Kong. On 25 February 1995, Nielsen was a guest at Bangkok's Alexander Hotel. Two days later, he purchased a Thai Airways ticket and on 5 March 1995, he yet again purchased a Thai Airways ticket. On 9 March 1995, Nielsen purchased a Cathay Pacific air ticket in Kowloon, Hong Kong, for Japan where, in Osaka, he made a payment at the Osaka Daiichi Hotel on 23 March 1995. His next destination was Taipei, Taiwan, where he purchased a Singapore Airlines ticket on 2 April, and then he was back in Osaka where he put up at the Hilton International Hotel. He paid for his room tariff there on 4 April 1995. He then flew back to Tokyo where he stayed at the Kiyaseikokuu Hotel and used the Visa card to pay for the room on 9 April 1995. Within three days Nielsen was in Copenhagen where, at the Plaza/Nissen Hotel, he made a payment on 12 April 1995. Two days later, he was in Laren, Holland, where he purchased a phone card, Fax o Fone BV, on 14 April 1995. The same day, he was in Amsterdam's SAS Royal Hotel. The next Visa transaction occurred at Osaka's Hilton International Hotel on 10 May 1995. A week later, he was in Lysaker, Norway, where he stayed at the SAS Royal Hotel, making a payment on 17 May 1995. He took a Swissair flight, which he paid for on 15 June in Copenhagen. On 21 June 1995, he was in Johannesburg, where he stayed at the Imperial Mooi Hotel and made a payment there on 29 June 1995.

There is, of course, no evidence to indicate who he met in the course of these fly-by-night visits to diverse countries, but he continued to move swiftly in July. He was in Monaco where he paid for food using his Visa card on 3 July 1995 at La Strega, a fast food restaurant at Rue de la Colle. The same day, he made a payment at AGIP, a gas station at Chatillon in Italy. He seems to have driven back to Monaco where he paid at (possibly) a restaurant by the name of La Siecle. The next day—4 July 1995—he drove back to Vezzi Portio where he filled up gas. The next day he was in Zurich where he paid at a Hertz Rent-a-Car outlet on 5 July. On 18 July 1995, he made a payment in one of Copenhagen's Hertz Rent-a-Cars. The next payment was noticed on the same day in Newark, New Jersey, where he picked up a car on rent to drive around. It was at Newark airport that fingerprints lifted from a Customs declaration form matched those taken by the Horsholm police in Denmark on 4 August 1978. The fingerprint comparison was made by the Forensics Document Laboratory of the US Immigration and Naturalization Service. As far as the ACIU and the CBI were concerned, Nielsen's real identity had been established.

On 20 July 1995, Nielsen made a payment at Hotel Lexington in the heart of Manhattan, New York. The same day he took a flight to Denver, Colorado, where he had a large and luxurious house in a ski resort area. He next flew to Los Angeles where he paid for a room at DoubleTree Hotel on 26 July 1995. He possibly met with Wai Hong Mak, who had already entered the US via Seattle before proceeding to Anaheim in California. (Mak left the US on 24 July 1995.) On 31 July, Nielsen made a payment at an airport near Parklands, a suburb of Cape Town. After that, he was back in Johannesburg where, on 3 August, he paid using his Visa card at the Imperial Mooi Hotel. His next destination was noticed to be Copenhagen again, where he paid for renting a car at a Hertz rental on 19 August 1995, a day after the meeting at Peter Haestrup's Klampenborg house. Between 3 and 8

October 1995, Nielsen travelled to Pakistan, sneaked into India unchallenged and then flew to Colombo, Sri Lanka, before reaching Seoul on 9 October. Neither the CBI nor Interpol nor Indian intelligence had any clue as to who he met with in Pakistan (the second visit to that country in seven months) or Sri Lanka during his early October sojourn to the two South Asian countries. This was one of the many loose ends that did not attract the attention of the CBI's investigators and remained unresolved, primarily because of lack of Pakistani cooperation with the Indian law enforcement agency.

I have already referred to all his recorded Visa card transactions since the Copenhagen meeting and will therefore not list them here, save for the ones made at Hotel India, Varanasi, where he paid $400 on 26 November 1995, and at Hotel Taj Ganges on 24 November. This was the 'dry run' for the AN-26 before Nielsen and the Latvian crew finally returned to Plovdiv in Bulgaria. Nielsen closed his Citibank Visa account (card no. 4910-1715-4336-7601) in May 1996. He possessed a second Citibank Visa credit card (no. 4568-8170-2002-7515) which he used very sparingly. The principal holder of the second Visa card was one Peter Brook (Australian passport no. 2081849) who lived in Raffles City, Singapore. Brook did not use the card after October 1995.

After escaping from India into Nepal, the first report of a Nielsen sighting took place on 29 January 1996, when a CBI source reported to the agency that the Dane had been spotted at an Ananda Marga establishment in Örebro, Gjuthuset, Sweden. The physical verification of the place and surveillance over a few days turned up a negative result. Two years from that date, reports emerged of more regular and 'authentic sightings' across the globe. A source, codenamed S82-84-85, reported on 15 January 1998 that Nielsen was in São Paulo in the company of Robert Caring, alias Robert Dean Child. According to this information, Child attempted to obtain three new identities of persons born between 1961 and 1963 but being born in 1953, he could not possibly pass for being born in 1961, as Nielsen was. A contact was established between the ACIU and the police authorities in São Paulo, who later managed to locate Caring's residence in the Brazilian city. The São Paulo police wanted to know if an extradition treaty between Brazil and India existed in case of a possible arrest. The contacts were on an extremely discreet level in order to avoid possible leaks. However, due to an unfortunate misunderstanding between the officers involved, an official message was sent on 14 March 1998 from Interpol, New Delhi, to Interpol, Brasilia, seeking Nielsen's arrest. After that incident, neither the ACIU nor the CBI ever heard from their Brazilian contacts again. At that time, Child was the subject of an investigation by the American Drug Enforcement Administration (DEA) and was wanted on a national level within the US. An extensive probe of Nielsen's telephone records and inquiries on his associates in some South American countries revealed his links with one Sergio Goldenberg of De Mayo Street in the Paraguayan capital of Asuncion.[5] A currency trader, Goldenberg was also suspected to be part of a cocaine trafficking organization with links in Colombia and a few other narcotics-producing South American countries.

Following the replacement of an Interpol Red Corner Notice against Davy by one against Nielsen on 16 January 1998, the US State Department shared information on 6 April 1998 with Interpol and CBI that 'according to a tip-off, Robert Caring was likely in Ecuador staying in a village named Guayaquil' and that Nielsen was possibly with him.[6] An alert was sent out, addressed to countries on the South American continent, informing them about the case and requesting them to update their border control authorities. On 15 April 1998, the Interpol unit in Bogota, Colombia, sent information on the movements of Nielsen, Robert Dean Child and one Robert Marshal Newman, who repeatedly travelled between Miami and a Colombian city, El Dorado. It was an extraordinary coincidence to have the three names in the same message. While Caring's alias, Robert Dean Child, was known, the information related to Robert Marshal Newman suggested he was an Ananda Marga activist in charge of the Proutist Universal office in Taipei, Taiwan, with a separate office in Hong Kong. However, the US State Department and Interpol Wiesbaden concluded that the three were 'most probably not our men'[7]. An even more remarkable piece of information, bordering on the Ludlumesque, was one which the US State Department shared with Brannerud's ACIU on 7 October 1998, that 'Nielsen was so concerned about getting caught that he underwent a plastic surgery in Italy along with another member of the Ananda Marga'[8]. Obviously, the information supplied could not be independently verified by either the CBI or the ACIU.

More credible was a piece of information, this time from an Interpol source, claiming that there was reason to believe that Nielsen had stayed in Sudan. Incidentally, Interpol Khartoum also provided the ACIU with information mostly based on the Red Corner Notices issued against Nielsen, Deepak and Tadbhavananda, alias Lal Chand Parihar. An even more definite report from Interpol's Khartoum-based National Centre Bureau (NCB) claimed on 25 November 1998 that 'Nielsen most probably has been located'[9]. Besides, the Sudanese security services, who also assisted Interpol, informed via NCB, Khartoum, that 'Nielsen stays in the southern part of the country'[10]. Less than two months later, on 6 January 1999, an American source codenamed S175 wrote to the ACIU, 'It is very plausible that N would be in southern Sudan at this time, waiting for his identity to cool down and to recoup financially.'[11] On 9 May 1999, ACIU received another report that Nielsen possibly attended the Ananda Marga's Qahira (Cairo) sectoral conference in Veli Losinj, Croatia.

But what most interested Interpol and CBI investigators was the presence of Nielsen in the US in 1998. Barely six weeks after Nielsen's 22 December 1995 escape, Bill Loose, a long-time Denver-based associate of Nielsen, filed a hand-written warranty deed at the Las Animas county (Colorado) land office, transferring the Dane's large farm property to White Winds Trust with a post office box address in Beverly Hills. This, investigators suspected, was done under the instructions of Nielsen, who was wanted in the US on two counts of passport forgery and counterfeiting $100 bills. On 29 June 1999, the sheriff's deputy in Trinidad city, Las Animas county, identified Nielsen from a photograph shown to him as a subject he had questioned the previous year in the vicinity of the farm. Nielsen had claimed he was backpacking in the area, but the deputy noted that the place was not known for trekking or backpacking. Repeated efforts on the part of Interpol USA to question a US-based uncle of Nielsen failed as he refused to cooperate.

Nielsen finally surfaced in Denmark in late 1999 or early 2000. He took up a rented house in a Copenhagen suburb, but was recognized by a retired prison officer who quietly reported him to the police which, inexplicably, did not arrest him, even though a valid Interpol Red Corner Notice was pending execution against him. In December 2002, Nielsen was photographed by a camera at a traffic intersection and spotted at a bus station. By that time he had turned bold enough to appear publicly in an interview to a Danish television news channel in the middle of 2000, when he denied that he was involved in the arms drop, a claim that was not believed by the Danish media, which continued to run a series of stories on the gunrunner, describing him as a terrorist. By that time, he was Number Two among the world's six most wanted criminals/terrorists.

Joel Broeren, alias Martin Konrad Schneider, alias Ken Sendo.

The Curious Case of Joel Broeren

As the relentless search for Nielsen continued, a curious event in the Purulia arms drop episode occurred with the sudden and dramatic arrival of an American national, Joel Broeren alias Martin Konrad Schneider, on a Singapore Airlines flight to Mumbai on 18 January 1997. The Mumbai police arrested the tall, well-built and bearded American of German descent as soon as he deplaned at Sahar International Airport. A senior functionary of the Ananda Marga's Seva Dharma Mission, Broeren, born at Thorp, Wisconsin, on 8 February 1954, spoke fluent Japanese, Mandarin Chinese, Korean, French, Sanskrit, Bengali and Hindi.[12] During interrogation by the Mumbai police and subsequently by the CBI— which brought him to Calcutta on a transit remand, produced him at the Bankshal Court and took him into custody (on the charges of violation of the Foreigners Act, counterfeiting/forgery, people smuggling and trafficking[13])—Broeren admitted to knowing Davy 'very well'[14]. A few months after the arms drop, the CBI was able to establish a link between Nielsen and Broeren, but it did not find any deep involvement on the part of the American.

During interrogation, Broeren also confessed that he first met Nielsen at a yoga seminar in Taiwan in the autumn of 1990. Towards the end of 1991, less than a year after the end of the Cold War, he again met Nielsen at Khabarovsk in Far East Russia. The Dane was then organizing a visit of some Proutists from the US, besides conducting seminars in Khabarovsk and Vladivostok, where he did business in Bridgestone tyres, old office furniture, used Japanese cars and Canon copiers. It was at Khabarovsk that Broeren helped Nielsen set up his first email account with the user ID 'DNA'. For two years, beginning at the end of the eighties, Nielsen lived in Khabarovsk which, during the Soviet era, was a city closed even to Russians without proper travel documents. At that time, he had built a large and effective network of contacts in Russia, one of whom was Mikhail Ivanovich Kalyukin alias 'Misha'.

In the early nineties, Nielsen and Broeren travelled together to Mumbai, where the American had business interests, and New Delhi, on several occasions. Between 7 and 12 January 1996, Broeren visited Seoul and Hong Kong, where he got access to Nielsen's Patterson Street apartment. Broeren disclosed that he knew Nielsen's assistant, Angel A. Caparaz alias King, a Filipino who lived in the Dane's Hong Kong apartment. Caparaz, according to Broeren, had earlier founded a company called Kultura Asia, which manufactured shoes in the Philippines. Later he set up Kultura Asia Pvt. Ltd in Singapore. Both the companies wound up after the arms drop, but Caparaz continued to be a devoted member of the Ananda Marga. In early January 1996, Broeren had an opportunity to speak to King who told him then that Nielsen had called him once on his landline phone after the arms drop, but had not disclosed where he was calling from. Caparaz left Hong Kong for the Philippines sometime in June 1997 and has never returned to the Chinese special administrative region.

Broeren admitted to CBI and R&AW interrogators that he also knew Deepak, alias Daya M. Anand, whom he had first met in Singapore in 1994. The two would often visit each other and travelled

together for the Ananda Marga Dharma Maha Sammelan at Ananda
Nagar in Purulia in December 1994. In his interrogation, Broeren
revealed that Deepak last visited him in Singapore on 21 or 22
December 1995. 'Deepak appeared quite nervous but did not go into
the details [of the arms drop],' Broeren told his interrogators in
Calcutta.[15] Confronted with a series of emails retrieved from Nielsen's
briefcase that he had left behind in the aircraft at Mumbai airport,
Broeren revealed the smuggling enterprise of some fellow Ananda
Margis, including expeditions to Ghana for extracting gold dust in
the West African country. Expressing surprise about Nielsen's arms
drop mission, Broeren told his interrogators that he was 'shocked
because of the brazen stupidity of such a move which had no strategic
or logistic value for our organisation in Purulia or elsewhere. Yet, I
was not "surprised" as this extreme risk-taking clandestine adventurism
certainly fits with his personality'[16]. Broeren claimed that between 7
and 12 January 1996, he visited Nielsen's Hong Kong residence on
Patterson Street to 'wash clothes'. Between 3 April and 2 May, he was
in San Francisco and then travelled to Colorado (possibly to Nielsen's
ranch in Las Animas).

It was not clear why, suddenly out of nowhere, Broeren, who held
fourteen different passports and used at least six other identities,
including Ken Sendo, Roy Dogen, Kurt Martin Olsen and Acharya
Krishnabuddhyananda Avadhoot, landed in India in the winter of
1997. He had last visited India in August 1995 to attend an exhibition
at the World Trade Centre in Mumbai as a representative of Vistara
Media International Pvt. Ltd, based in Singapore. It appeared that
his inexplicable 1997 visit to India had a dual mission: to mediate
between two factions in the Ananda Marga (one of which blamed the
other for having planned and assisted the Dane in the arms drop) and
to assist the CBI with information on Nielsen. The anxiety to steer
clear of any controversy was to avoid a second possible proscription of
the Ananda Marga by the Indian government.

Three years after his sudden appearance in India, Broeren as
suddenly disappeared, giving the CBI the slip. After being granted
conditional bail by the Calcutta High Court on 24 April 1997, he was
forbidden to leave Calcutta without the court's permission. According

to the terms of the bail, Broeren was required to appear before the investigating officer thrice every week. On 21 January 1999, Broeren's lawyer submitted a petition in the high court, praying for relaxation in the bail terms so that Broeren could visit Wisconsin in the US following the death of his father. The lawyer submitted that his mother had asked him to return home to attend the funeral. Moreover, his petition said, he was to take charge of the family property. The court allowed him to visit Wisconsin on humanitarian grounds. Broeren had to furnish two sureties of Rs 1 lakh and the court directed him to make periodic appearances before the Twelfth Metropolitan Magistrate at the Calcutta Sessions Court. It was during routine perusal of documents relating to the arms drop case that CBI officers in Calcutta realized, after almost a week had passed, that Broeren had not turned up. The magistrate told his lawyer that Schneider's (Broeren) stay in the US could not be extended. An arrest warrant was issued and the CBI 'revived' the look-out notice which had been dispatched through Interpol when Broeren's name first came up in the case.[17] Needless to say, Broeren never returned to India. After spending some time in Wisconsin, Broeren switched his name to Ken Sendo before moving to California, where he launched a company called Isagenix International that promotes anti-ageing drugs and solutions. He maintains discreet contact with Nielsen via Facebook through intermediaries in the Ananda Marga, including Nielsen's so-called assistant/associate in the mid-nineties, King.

Who is Niels Christian Nielsen?

What emerged from various sources was the profile of a man whose mind functioned like a seasoned and well-trained intelligence operative. According to Peter Bleach, who worked with Nielsen during the four-month-long arms deal-and-drop project, even though 'a really good overall plan was spoilt by really bad attention to detail', Nielsen was a 'very level-headed person'[18] and one who would not panic easily in the circumstances that he was faced with—when the plane was asked to land at Sahar International Airport, when confronted with the situation at the airport and after he escaped. A question that arose in Bleach's mind after his capture was why, considering he was a crafty operative, Nielsen had left behind his briefcase (containing documents that could potentially incriminate him) in the aircraft just before escaping. In his typed tell-all statement to the CBI, Bleach wrote,

> There has to be a possibility that he left it behind because he was already consciously abandoning that identity [of Kim Davy] and he knew that there was nothing in it either that could really help the investigation. He had already been through the briefcase very carefully before we landed and had burnt those documents which he did not want found.[19]

For all the chase and the global manhunt against Nielsen, the CBI and Interpol possess very little information on the Danish gunrunner. Born into a middle-class Danish family in the small town of Aars, in the Himmerland area of Jutland, Denmark, on (curiously) 17 December 1961 (although the passport in the name of Kim P. Davy showed he was born in 1962), Nielsen was an early school dropout and took to petty crimes late in his teens. He first came to the notice of the Danish police in the late seventies—1978 to be precise—when he was accused of robbing a supermarket in Copenhagen. He was arrested by Horsholm police on 4 August 1978, but managed to escape from the court premises, where he had been brought to stand trial, in his socks, an act that earned him the sobriquet 'Barfodsrøveren'[20] or 'barefoot robber' in Denmark. Soon afterwards, in 1982, he was suspected to be involved in a similar robbery in a jeweller's store in Gothenburg, Sweden. Nothing at all is known about Nielsen in the years since he eluded capture by the Danish and Swedish police till 1985 when, it was believed, he joined the Ananda Marga. Working for this shadowy right-wing Indian cult organization took him to countries in South America, east Asia and Russia where, in the heavily militarized city of Khabarovsk, on the eastern edge of the country, he lived under cover for two years, apparently conducting Ananda Marga yoga seminars, Proutist Universal meetings and conferences, and in the process, developing wide-ranging contacts among native Russians.

In the early nineties, Nielsen appeared to have shifted base to

Hong Kong where, in February 1991, he, along with a few other American and British nationals, established a company called Howerstoke Trading Ltd. Its offices were at Kinwick Centre, 32, Hollywood Row, Hong Kong.[21] While 'working' for Howerstoke Trading Ltd, Nielsen used the alias Peter Johnson. Among some of the businesses the company was involved in at the time was gold trading, which became Nielsen's main business/smuggling activity.

While 'Peter Johnson' was also suspected to be involved in smuggling gems and other precious stones from South Africa and Sudan, his main interest continued to be in the southern Sudanese gold mines, especially in Kapoeta in Eastern Equatoria province and Jonglei state. In an undated fax message (possibly sometime between the end of July 1995 and the first week of August 1995) to one Ivan Whitehead in Johannesburg, 'Peter Johnson' inquired about mining prospects in Sudan and suggested sending 'a couple of specialists to survey the area and make a feasibility study for the whole operation'[22]. Regrettably, neither the CBI nor Interpol could investigate the scale and magnitude of Nielsen's nexus with the SPLA and the exact role he played in the weapons-for-gold operations in the early and mid-nineties. Diplomatic channels were activated with Khartoum to track down Nielsen in an effort to find out the extent of his involvement with the SPLA, but the Sudanese government under Omar Hassan Ahmed Al-Bashir did not respond favourably. 'There were no reliable partners who could communicate with Sudanese officials,' an Interpol source said.

What the two law enforcement agencies could, however, gather in great detail was that Nielsen or the suspected financier of the arms and ammunition and the aircraft, Wai Hong Mak, had deep pockets. Nielsen had access to very large sums of cash in US dollars, which would be delivered to him by courier (or by Mak personally, as happened at Phuket where the Chinese travelled to on 19 December 1995 to deliver $20,000 in cash to Nielsen) at any place in the world within forty-eight hours' notice. A document the CBI extracted from Nielsen's laptop after it was seized from the AN-26 provided details of the money spent against various heads, including air tickets for a 'Mr Schumann', machines, house and 'sample stones' (suspected to be precious stones, including gems) in South Africa, as well as a summary of expenses incurred in Sudan and Italy. Since the document was addressed to Mak, the CBI suspected that he might have provided capital and had interests in Nielsen's projects in Sudan. If Nielsen was an Ananda Marga follower—which undoubtedly he was—his spiritual inclinations belied his smuggling and other covert activities. There was more to his Ananda Marga activities than met the eye.

Indian investigators and security agencies familiar with the Purulia arms drop case are aware of only five aliases (Kim Palgrave Davy, Kim Peter Davy, Neil Davy, Peter Johnson, Dada Nirvanananda and Dada Anindyananda) and two passports with non-sequential numbers (nos. X-163495 and Y-341624, issued on 13 September 1992 and 30 September 1991 respectively, in Wellington, New Zealand)[23] that Nielsen used. One of the New Zealand passports was acquired on the basis of a certificate furnished by a US citizen, David Mathews alias Mathew David Meighan (American passport no. 1443361), who was born in San Francisco in October 1953. David Mathews, who had tried to enter Australia from Thailand in 1981 as a 'guerrilla trainer', figures in the Indian Ministry of Home Affairs' Suspect Index, 1996.[24] It is estimated that Nielsen might have used as many as thirty-six aliases and fifteen passports in a span of ten years. It is a measure of his extraordinary skill in assuming false identities, including that of an Indian (passport no. P4735880, issued in New Delhi)[25], that though born a Dane, he was almost always successful since 1982 in passing himself off variously as a New Zealander, a Dutch and several other nationalities. He is a polyglot, proficient in at least eight languages—Danish, English, Russian, German, French, Canton Chinese, Hindi and Bengali—involving five alphabets, suggesting that he is either a natural linguist or was trained to be one. Although Nielsen preferred to operate on his own—relying on stealth, his own keen senses and street-smart intelligence—he did have the backing of a large number of non-Indian Ananda Margis, some of whom were suspected to have helped him cross over to Nepal and then to find safe sanctuary in some South American countries and finally in Denmark, where he first surfaced in late 1999.

Nielsen would move about openly on the streets of Copenhagen, giving interviews to local Danish and British newspapers and even to foreign television news channels, justifying the arms drop as an attempt to arm the Ananda Marga threatened by the communists in West Bengal, besides proclaiming his innocence on the ground that he was merely driven by an ideal to assist the oppressed people in Purulia so they could defend themselves from state repression. In 2008, he co-authored a book *De Kalder Mig Terrorist* (*They Call Me a Terrorist*) with a Danish journalist, Øjvind Kyrø, indicating, quite nonchalantly if not arrogantly, that the arms drop was the work of Indian intelligence whose operatives helped him escape from Mumbai. His book and its contents were not taken seriously by a majority of his countrymen.[26] Four years before his book got published, Niels Christian Nielsen married Silvia Holck and adopted her surname (which is rare but not unusual in Denmark)[27], giving himself yet another identity. He fathered two children, one in 2004 and the other in 2007. On 5 October 2009, in an act of defiance, mocking the criminal justice system not only of India but also Denmark, Niels Holck opened a Facebook account. Today, the man against whom an Interpol Red Corner Notice is pending execution, merrily posts away on his Facebook page. However, a change in name and settling down to a new life and a new business—a company called Living Waters manufacturing and exporting houseboats to other parts of Europe—did not quite help him shrug off the arms drop episode.

Attempts to Proceed Against Haestrup, Thune

When the CBI approached the Danish authorities in August 1996, seeking their assistance in investigating the Purulia arms drop involving Peter Jorgen Haestrup and Brian Thune—Nielsen had not been officially identified as a Danish national at the time—little did it know of Copenhagen's extreme reluctance to cooperate in the probe. That reluctance would turn into deep distrust of the CBI and almost complete refusal to take any criminal action against Nielsen, despite a valid Interpol Red Corner Notice against one of the most wanted men in the world, after he surfaced in Denmark towards the end of 1999 or beginning of 2000. 'We were absolutely unprepared for the inexplicable feet-dragging and inertia on the part of the Copenhagen authorities, including the country's Criminal Investigation Department and the Ministry of Justice, that we encountered after Nielsen surfaced. We could sense a degree of indifference,' a senior CBI officer who was then involved with the investigation but has since retired from service, told me, requesting anonymity. Nevertheless, with Interpol's support, the CBI tried to engage the Danish police so that criminal action could be taken against Haestrup, Thune and Nielsen.

When the CBI first got in touch with the Danish authorities in August 1996, it requested, inter alia, carrying out inquiries against Haestrup and Thune. It took the Danish Ministry of Justice eight months to respond to the Indian request, but when they did so in May 1997, they said that while the two Danes had been identified, they would prefer to know whether the CBI had charged or intended to charge Haestrup and Thune. The Danes sought to know the offence with which the two had been or would be charged and the potential sentence that might be imposed on them. The Danish authorities dispatched two more letters, on 31 July 1997 and 19 February 1998, repeating their position. Subsequently, at a meeting between representatives of the Danish Ministry of Justice, the national commissioner of police, the commissioner of Copenhagen Police, Indian Embassy officials and the CBI on 14 September 1999, the Danes placed three conditions under which they would proceed criminally against Haestrup and Thune. These were:

1) The Indian authorities must agree that regardless of the results of the interrogation/investigation in respect of the two Danish citizens, according to Danish law, they would not seek Haestrup and Thune's extradition.

2) Should the CBI find that the interrogation reports of Haestrup and Thune provided reasons for charging them, the Indians must agree that the Danish police take over criminal proceedings against the two Danish citizens and in this connection transfer the relevant evidence against them to the competent Danish authorities.

3) Given the fact that the two Danish citizens were suspects in the case and, therefore, according to Danish law, were not obliged

to tell the truth, the Danish police's interrogation/investigation reports would not in themselves be considered as evidence.

The Indian representatives in the meeting said that they would consider the Danish proposals. Following the return of the CBI officers to India, a meeting of the top brass, held at the agency's Lodhi Road headquarters, decided to comply with the Danish suggestions and sent out a written assurance, agreeing to all of the conditions. Based on the CBI's written assurance of 11 March 2000, the Danish CID shared (in May 2000) reports of the interrogation, which had already been completed between September and October 1996, after receipt of the Letters Rogatory by the Chief Metropolitan Magistrate's Court in Calcutta on 15 July 1996 reached the Danish Ministry of Justice.[28] The Danes wrote back to the CBI, through the Indian Embassy in Copenhagen, 'Based on the information available at this time, it not possible for the Danish Ministry of Justice to assess which provisions of Danish law would be applicable if criminal proceedings were to be conducted against the Danish nationals in Denmark.'[29] Besides, the Ministry of Justice said that since documents were 'not admissible as direct evidence in a trial if it comes to criminal proceedings...[it] would be necessary for relevant witnesses to testify directly before a Danish court of law.'[30]

When the CBI sought more information on Haestrup and Thune, the Danes again tried to stonewall not just the pace of likely criminal proceedings, but also created a maze of bureaucratic red tape, imposing more conditions. The Danish Ministry of Justice responded on 22 January 2001, saying,

> The Ministry of Justice will forward the request (for the Danes to take over criminal proceedings against Haestrup and Thune) to the relevant Chief Constable/District Prosecutor in Denmark who will, according to Danish law, will [sic] have to consider whether it is possible for Denmark to take over proceedings. The Chief Constable/District Prosecutor will also consider what evidence, which [sic] will have to be submitted to the Danish authorities, and which witnesses that will have to appear before a Danish court.[31]

For all practical purposes, the Danish Ministry of Justice had effectively stalled the CBI's attempts to even have Haestrup, the son of an influential businessman, and Thune tried in Denmark. The Ministry of Justice said, 'If proceedings are transferred to Denmark and witnesses have to be heard before a Danish court, the witnesses will have to travel to Denmark and give statement[s] directly before the Danish court,' and 'It is not possible at this stage to estimate whether Peter Bleach will have to give statement in Denmark in case proceedings might be initiated... It will be the Chief Constable/ District Prosecutor who will have to consider whether Peter Bleach has to give testimony in Denmark.'[32] It did not even entertain an Indian proposal that a two-member CBI team visit Copenhagen to discuss 'various points related to the case' before an Indian request for transfer of criminal proceedings was made. An extensive investigation into Haestrup and Thune's roles, leave alone a trial in a Danish court, never took off. The underlying basis for Danish non-cooperation at the most and reticence at the least was reflected in then Danish Justice Minister Lene Esperson's response to a question in Parliament in 1999, where she said that,

> The Ministry has not found it correct to contribute to investigation against Danish citizens as it does not appear with sufficient clarity what the Danish investigative material will be used for, and if it may result in the Danish citizens in question risk being sentenced to penal sanctions on the basis of the Danish investigative material which will be incompatible with fundamental Danish judicial principles such as capital punishment.[33]

Whatever the Danish stand, Interpol's views on Haestrup were clear:

> Haestrup seems clearly to be involved in the case at an early stage when recruiting Peter Bleach for this mission... He participated in the preparations in order to complete the [ams] and aircraft purchase. He arranged the first meeting in his residence in Copenhagen. He participated in the meeting at the Swiss Felix Hotel in Bangkok after having travelled to Bangladesh together with Bleach.[34]

At the CBI headquarters, there was gloom and doom. Seniors officers, stung by Copenhagen's frosty attitude, became concerned that they would face bigger walls in their pursuit of Nielsen, who had by that time already surfaced in Copenhagen. Meanwhile, the chief Interpol representative in Denmark at the time, Palle Biehl, informally advised his Indian counterparts in New Delhi to concentrate on Nielsen and make an all-out effort to have him extradited from Denmark to India. At the same time, a review by some CBI officers of the case against Haestrup and Thune suggested that there was no corroborative evidence against the two, even though Peter Bleach had admitted to and provided details of the role played by them in the conspiracy. During an interview in New Delhi in October 2012, a CBI officer told me,

> We considered that Bleach's statement, that Haestrup attended the 27 September 1995 Bangkok meeting in which the place where the weapons would be dropped was decided, among other things, did not amount to evidence that could be supported or corroborated by other independent evidence or witnesses. Even the documentary evidence that originated from Thailand, in response to the Letters Rogatory sent to the Thai authorities, would not amount much for evidence as the Danes had already declared that investigation documents would not be admissible as evidence in a Danish court of law.

The CBI was not interested in pursuing William Roeschke, the Munich-based Danish businessman who had recommended Bleach's name to Peter Haestrup. Sometime in the spring of 1996, Roeschke was questioned by the Bundesnachrichtendienst (Federal Intelligence Service) which did not pursue the matter further because his role was found to be minimal. Roeschke was also examined by the Danish Politiets Efterretningstjeneste (PET) or the Danish Security and Intelligence Service, which too did not press the issue further. Today, Roeschke, who runs a company that deals in aeroplane engine turbines, is 'not very proud' of his involvement in Bleach's proposal to procure an aircraft, which the Briton said would be used for transporting tiger shrimps from Pakistan to Hong Kong. He recalls,

It was a dark chapter in my life and I want to forget it. Bleach wanted to know if I could supply drop-parachutes. On top of everything, he did not even give me the promised $20,000 commission if I could locate a suitable aircraft. In those days, I was low on money and the sale of the aircraft would have been my lifeline... Thinking back and from accounts I have heard from my sources in Denmark, it seems that Niels Christian Nielsen had friends in a foreign agency that may have influenced the Danish authorities to protect him. Nielsen was a front for someone else's operation.[35]

Going after Nielsen

Regardless of the reverses the CBI faced in the case of Haestrup and Thune, the agency began pursuing Nielsen's extradition to India in 2003 when Loknath Behera, who had now been promoted to the rank of deputy inspector general, visited Copenhagen. Pressure mounted at home as political parties and the media focused on the CBI's apparent inability to bring Nielsen, the fugitive from law, to justice in India. Although there was no extradition treaty between Denmark and India, Behera met the who's who of the Danish security, police and Ministry of Justice establishments, impressing upon them the absolute necessity to extradite Nielsen, especially because they were duty-bound to detain him and hand him over to the CBI in accordance with the rules and regulations that governed Interpol Red Corner Notices. It would be a tall order and, as subsequent events would show, absolutely impossible.

In the initial exchange between the CBI and the Danish police and Ministry of Justice over Nielsen's extradition, the Danish authorities '...in the period between 19 April and 14 May 2002 notified/ informed the Indian authorities that Niels Holck [he had changed his name by adopting his wife's surname Holck in 2004] could not be extradited for criminal prosecution in India. This was in accordance with the extradition acts applicable at that time pursuant to which Danish citizens could not be extradited for prosecution in countries outside the Nordic region.'[36]

Subsequently, especially after amendments were made to criminal legislation, including the law on extradition, following the 11 September 2001 terror strikes in the United States, the Danish authorities took an in-principle decision to make a departure from Nordic law on extraditing their citizens to other countries to stand trial. The Indian authorities, therefore, formally sought Nielsen's extradition to India on 23 December 2002.

There was some unanimity among the Danes that the Purulia arms drop was a case of international terrorism. However, the Danes put forward a dozen conditions, each more stringent than the ones applied in the case of Haestrup and Thune, to the CBI. The Indian investigating agency even agreed to the condition that if Nielsen was convicted in an Indian court, he would serve the sentence in a Danish prison. The Indian authorities assured the Danes that the Indian cabinet was empowered to take a decision on transferring Nielsen to a prison in Denmark, an action which Prime Minister Manmohan Singh's cabinet did indeed take in 2006. A 'sovereign assurance,' which specified in unambiguous terms that Nielsen would not be awarded the death sentence, that his trial would be conducted expeditiously on a day-to-day basis and that all his human and civil rights would be protected, was sent to the Danish authorities.

When the Indian extradition order was brought to court for adjudication by the Danish Ministry of Justice, the City Court of Copenhagen appointed Niels Frosby as counsel for the defence of Nielsen on 28 February 2003. In a written communiqué to the Danish Ministry of Justice on 28 March 2003, Frosby wrote that his client Niels Holck admitted to 'having helped with the smuggling of small arms (AK-47 rifles) to the village in Purulia on 17-18 December 1995,' but denied 'Holck having been involved in ordering rocket equipment, grenades, sniper rifles and mines.'[37] Thereafter, between 2004 and 2007, a flurry of letters and other written communication related to Nielsen's extradition were exchanged between the Danish Ministry of Justice, the police and the CBI. These were primarily related to the assurances given by the Indian government as well as the CBI on humane treatment to Nielsen, prison conditions in India and dilution in the charges initially brought against him, including the removal of the Indian Penal Code provision on 'waging war against the state' and its corresponding clause on the death sentence. On 14 September 2007, the City Court of Copenhagen appointed Tyge Trier as Nielsen's new counsel. Three months later, Trier wrote to the Ministry of Justice to separate the legal basis for Nielsen's extradition for 'separate treatment', a move that was rejected by the ministry.

In fact, the Danish Ministry of Justice now moved with resolve, or appeared to do so, on Nielsen's extradition. After examining the material evidence of the case, it found that 'conditions for extradition...are fulfilled with regard to extradition for criminal prosecution for violation of the provisions of the Indian Penal Code, the Indian Arms Act and the Explosives Act.' In particular, the Ministry of Justice concluded, after an 'overall assessment, including seem [sic] in [the] light of the fact that [the] Indian authorities have confirmed to abide by the terms [and conditions]' it did not find 'any ground for denial of extraction with reference to Niels Holck not expecting to get fair trial and public hearing within a reasonable period or with reference to him being subjected to torture or other inhuman or degrading treatment.' Besides, the ministry held that 'extradition would not be incompatible with humanitarian conditions.'[38]

Meanwhile, the Indian government began to apply pressure on the Danes to quickly extradite Nielsen. In February 2008, the then Danish prime minister, Anders Fogh Rasmussen, on a visit to New Delhi ahead of the forthcoming climate summit in Copenhagen, declared that Nielsen's extradition was very much on the cards, a statement which was reiterated by his successor, Lars Løkke Rasmussen, in September 2009 during his official visit to the Indian capital. By April 2010, then Prime Minister Lars Barfoed had taken a firm decision to extradite Nielsen to India where he would stand trial. As far as the Indian establishment, including the CBI, was concerned, this was good news. The arrest of Nielsen by the Copenhagen police on 9 April of the same year was heralded as a vindication of the investigating agency's efforts to bring to book one of the most wanted terrorists in the world. There was jubilation at the CBI New Delhi headquarters that finally, after a fifteen-year-long manhunt, Nielsen was days away from being tried in an Indian court of law. Officials handling the case at that time sounded confident that they would be able to 'break' Nielsen to reveal not only who planned the Purulia arms drop, but also to extract details on the ultimate end-users. Surprisingly, on 10 April, Nielsen was granted bail with the mild condition that he would keep the Danish police authorities informed of his availability in Copenhagen over telephone.

When Nielsen's extradition matter came up for hearing, the Hillerod Court, however, viewed the case, including Nielsen's involvement, differently. In a lengthy reasoning, which was based not so much on the incriminating and unimpeachable evidence against Nielsen as much as on questions of perceived inhumane treatment he might receive in an Indian jail, the court said:

> The court finds the contents of the defined terms to some extent lack sufficient precision and detail especially with regard to the requirements for the circumstances under which Niels Holck is likely to be detained in India.
>
> On the basis of the information stated, the court takes into account that administration and operation of the prison service and local police in India is a constituent state matter, and that in India there is no system or body that has the opportunity to conduct inspection independent of the Indian authorities as regards prison service and police.
>
> There is thus no opportunity for independent qualified persons to conduct efficient monitoring of compliance with the stated terms and conditions.
>
> The information available data [sic] on the human rights situation in India describe and document very significant problems with violence and human rights violations at the constituent state level, including in West Bengal. Based on the evidence available it is found unobjectionable to take into account that in India at constitutional state level there are considerable difficulties in meeting and observing internationally accepted standards and legal security guarantees in relation to detention and incarceration of suspects.

The court therefore finds that, notwithstanding the Indian Foreign Ministry's confirmation, it must be regarded as predominantly dubious to assert that the Indian central government efficiently and in a fully satisfactory manner will be able to ensure that all established terms and conditions for extradition will be observed in every respect by the local constitutional state authorities responsible for the imprisonment of Niels Holck after extradition.

In this context, emphasis must also be attributed to the contents and extremely serious nature of the charges raised against Niels Holck by the Indian authorities.

Upon a total evaluation, the court therefore [has] sufficient basis to conclude that there is such a real and present danger that Niels Holck, if he is handed over for prosecution in India, will be subjected to conditions and treatment in violation of Section 6 of the Act on Extradition.

The terms and conditions for extradition of Niels Holck for criminal prosecution in India have therefore not been met. Therefore, the Ministry of Justice's decision of April 9, 2010, can not [sic] be approved.[39]

As for the other bases on which Niels Holck's extradition was being sought, the Hillerod Court accepted the argument of the Danish Ministry of Justice. On the issue of the previous final decision of the Danish government, the court, which took into account that there was no extradition treaty between Denmark and India, said, 'It cannot be assumed that in connection with the information/notification a final binding decision was made which prevents the Ministry of Justice...from being able to make a final decision on the extradition of Niels Holck.'

On equal treatment and information about the case, the judgement said that

...on the basis of information available, the court finds that the other two Danish citizens [Peter Jorgen Haestrup and Brian Thune] involved have played a considerably less significant role in connection with the actions on the basis of which Niels Holck is now requested to be extradited for criminal prosecution. The court therefore finds that according to other information available, there is no reason to assume that in connection with the processing of the case, the Ministry of Justice has employed subjective considerations or treated Niels Holck in conflict with equal principles normally applied.

On the issue of dual criminality and sentencing requirement, after discussing the Danish laws on extradition as they existed in 1995 and following amendments to them in the backdrop of the 11 September 2001, terrorist attacks in the United States, the court declared that

> According to information obtained and Niels Holck's explanation, he was the active initiator of and participated in the dropping of the arms shipment on December 17-18, 1995. Niels Holck has explained in particular that in the process he got a feeling that the shipment could contain more than just small arms (AK-47 machine guns) and ammunition as the shipment with regard to the credibility of the utilised 'end-user certificate' should appear differently. According to Niels Holck's explanation, the purpose of the arms dropping was to arm 3x75 men in private vigilante groups to protect the Ananda Marga-related areas, villages and projects in the state of West Bengal against illegitimate attacks by government affiliated groups and government supporters.
>
> The court accepts that the action leading to charges of violation of India law on weapons, explosives and explosive substances were, and are, if corresponding actions were committed in Denmark, [be] punishable pursuant to the Act on Weapons and Explosive Substances.
>
> The court accepts the reasons stated by the Ministry of Justice that the maximum and minimum sentences for the *corresponding* actions at the time when decision on extradition is made can be applied. The sentencing requirement of Section 2(1) (ii) of the Act on Extradition has therefore also been

met with respect to the raised charges of violation of Indian law on weapons, explosives and explosive substances.

It must, according to the available information, be assumed that a foreign intelligence service (the British MI5) prior to the weapons air drops was aware that preparatory actions were made in terms of arms procurement, etc., though the details about the specific knowledge is left uncertain. Furthermore, it appears that in autumn of 1995 PET [the Danish Security and Intelligence Service] was made known of at least one of the Danish citizens' possible involvement in an arms deal.

Notwithstanding this fact, and regardless of Niels Holck's belief that there must have been some participation from Indian authorities, particularly in relation to an 'open window', the court finally accepts the reasons stated by the Ministry of Justice that special considerations for law enforcement advocate the request for extradition.[40]

On the issue of political crime and the UN convention on suppressing terrorist attacks and bombings, the Hillerod Court ruled that 'Given the extent and nature of the dropped weapons and regardless of Niels Holck's claimed purpose of the weapons delivery, the action is found rightly to be comprised by [the provisions] of the [UN] Convention for the Suppression of Terrorist Bombings... Irrespective of Niels Holck's action being deemed politically motivated, this does not, according to Section 5(3)(iii), preclude extradition for prosecution in India.'[41] In this context, the court observed that the arms drop case was not time-barred.

In rejecting Niels Holck's argument that he should not be extradited on the ground that he had a family, the court declared, 'Regardless of the long process and Niels Holck's current personal and family circumstances, the court accepts the reasons stated by the Ministry of Justice that extradition for prosecution in India would not be incompatible with humanitarian considerations pursuant to Section 7 of the Act on Extradition or in violation of Article 8 of the ECHR [European Commission on Human Rights].'[42]

Needless to say, the Hillerod Court's judgement came as a huge blow for the CBI, if not the Danish Ministry of Justice. The CBI did not cover itself in glory either. Before the Hillerod Court took up the extradition case, a representative of the so-called premier investigating agency reached Denmark with an expired warrant of arrest for Nielsen, leaving officers in New Delhi red-faced.[43] But worse was in store for the CBI, this time in the form of a judgement of a Danish High Court which the Danish Ministry of Justice had moved in appeal against the Hillerod lower court order refusing Nielsen's extradition to India. A five-judge, and not the usual three-judge, bench of the Eastern High Court, took a swift and unanimous decision on 30 June 2011, upholding the lower court's decision. Agreeing with all of the Hillerod court's order, the Eastern High Court declared, 'The High Court therefore finds that the Justice Ministry's decision of April 9, 2010 on the extradition of D [the court appeared to be so protective of Nielsen that it did not even name him in its judgment, using only the alphabet D to describe him] for prosecution in India [is] contrary to the provision of the [Danish] Extradition Act...and that extradition should be refused.'[44] It seemed that Nielsen was immune to any kind of legal measure.

What baffled representatives of the CBI was that both the Hillerod and the Eastern High Courts had allowed Nielsen to field witnesses, including Peter Bleach and Peter Haestrup, while the Indian agency was forbidden to produce witnesses. What perhaps swayed the High Court's decision was Bleach's testimony, in which he described the terrible hygienic and security conditions in Calcutta's Presidency Jail. Though Bleach admitted that Nielsen conspired to purchase the weapons and the aircraft and was instrumental in airdropping the arms, he narrated the manner in which the CBI had earlier promised to treat him leniently and later went ahead to prosecute him. CBI officials now rue that even though the High Court's ruling was a foregone conclusion, the Indian government could have sent to Copenhagen the country's attorney general, considering it was a high-profile case that needed sensitive and careful handling best not left to Danish law firms.

Nothing seemed to have worked for the CBI, not even the solemn pledge that only the Indian federal government—not the West Bengal government—could provide an assurance on humane treatment in jail. To make the promise binding, a senior CBI officer even met the West Bengal Home Secretary and took an assurance in writing that the state government would abide by any decision of the central government on Nielsen.

The CBI deputy inspector general, Arun Bothra, who attended all the hearings at the Hillerod and the Eastern High Courts, came away with the suspicion that 'everything appeared to have been fixed'. Describing Danish Chief Prosecutor Jorgen Steen Sorensen's refusal to produce witnesses as 'intriguing,' Bothra said,

> I pleaded with the chief prosecutor that I would personally depose as a witness; he did not pay any heed. I literally ran after him outside the High Court to persuade him to be not just proactive but to field me as a witness. I tried to pass on relevant documents which he could have produced before the court, but he simply would not take them. He practically did nothing. He did not even counter the defence.[45]

Did Bothra smell a rat? 'Well, everything appeared to have been pre-arranged, stage-managed. I felt that on the one hand the public posturing was to extradite him and on the other the prosecution did not appear to be very keen or clean,' the CBI officer lamented.[46] What reinforced his suspicion of institutional protection for Nielsen was that the Danish chief prosecutor did not give him any reason whatsoever why the Ministry of Justice had decided not to move the Supreme Court in appeal. 'The Danish institutions were lukewarm and cagey, to say the least,' an Indian diplomat posted in Copenhagen at the time said, requesting anonymity.

What possibly motivated the Danish Eastern High Court to take a favourable view on Nielsen was the fugitive's sensational, but outlandish and motivated, claim on the Indian television news channel Times Now that the Purulia arms drop was a joint operation of R&AW and the MI5 to overthrow the communist government in West Bengal.[47] 'Additionally, the Indian government's soft approach, pardoning the Latvian crew members and Peter Bleach, raised doubts in the minds of the Danish judges whether it was hugely exercised in proactively bringing a closure to the case,' a former Indian ambassador to Denmark, who did not want to be identified, said. In the same interview to Times Now, Nielsen claimed that an Indian politician in league with the Prime Minister's Office had ordered the weapons consignment for use against the Marxist regime. More about this and the Indian government's reaction to this charge in the next chapter.

India's hopes of ever trying Nielsen on its soil were dashed for ever when the Danish Ministry of Justice took the strange decision not to move the country's Supreme Court as a last resort. Although the Ministry of Justice had two weeks to take a decision to move the Supreme Court, Chief Prosecutor Jorgen Sorensen took the unilateral decision in less than a week, preferring not to contest the High Court's judgement. On 7 July 2011, in the presence of a battery of Danish and Indian journalists, Sorensen said in a statement:

> The Prosecution Service will not to seek permission to bring the question of extradition of Danish national Niels Holck to India before the Supreme Court. This decision was made by the Director of Public Prosecutions today (July 7, 2011). My decision means that the Eastern High Court's order not to extradite Niels Holck will stand... The District Court as well as the High Court ruled in favour of Niels Holck, finding that extradition to India would be contrary to section 6(2) of the Extradition Act. According to this provision, extradition may not take place if there is a risk that the subject in question will be exposed to torture or other inhuman or degrading treatment or punishment after extradition. This provision corresponds to Article 3 of the European Human Rights Convention.
>
> The case has raised a number of important questions concerned with the interpretation of the Extradition Act, and the courts have to a wide extent agreed to the assessments made by the Ministry of Justice. Thus, both the District Court and the High Court agreed for example that the evidentiary basis for extradition is sufficient, that the double criminality requirement of the Extradition Act is satisfied and that the case is not time barred. The Courts have found,

however, on the basis of a specific assessment of the conditions under which Niels Holck may be expected to be detained after prospective extradition to India, that there is a real risk that Niels Holck will be exposed to treatment that is contrary to Article 3 of the Human Rights Convention... Against this background I do not find that the questions involved in this case are of a nature that will justify an application for permission to bring it before the Supreme Court.[48]

Recalling Sorensen's decision, a CBI officer in New Delhi told me on 24 October 2012 that for the investigating agency 'it was an eyebrow-raising choice, especially because an appeal to the Supreme Court was our last resort. It was a choice which should have been exercised regardless of the circumstance and the likely outcome.' Sorensen's decision baffled even the Danish media, for whom this was one of the biggest stories in recent times. Several of my journalist friends in Copenhagen told me over telephone that they were surprised by the chief prosecutor's hasty decision which, they hoped, was not taken on the prompting, advice and suggestion of a third country that might have had an interest in protecting Nielsen, especially his real identity and who he worked for.

What was overlooked by the Danish chief prosecutor, willingly or otherwise, was Denmark's obligation as a signatory to several international conventions on terrorism and organized crime, to apply the principle of 'aut dedere aut judicare', which enjoins on the state where a person accused of committing an offence of an international nature is to be prosecuted in the event of a decision not to extradite him or her. The precise reading of the obligation stands as:

A state may not provide a safe haven for a person suspected of certain categories of crimes. Instead, it is *required* either to exercise jurisdiction (which would necessarily include universal jurisdiction in certain cases) over a person suspected of certain categories of crimes or to extradite the person to a state able and willing to do so or to surrender the person to an international criminal court with jurisdiction over the suspect and the crime...when the *aut dedere aut judicare* rule applies, the state where the suspect is found must ensure that its courts can exercise all possible forms of geographic jurisdiction, including universal jurisdiction, in those cases where it will not be in a position to extradite the suspect to another state or to surrender that person to an international criminal court.[49]

The CBI officer who attended the hearings at the Hillerod and Eastern High Courts too did not raise the issue, which was largely because of ignorance about international norms, conventions and obligations rather than any deliberate effort to go slow or do nothing at all. Besides, without the services of the Indian attorney general, the country's topmost law officer, the CBI had to rely, much to its frustration, wholly on the legal position taken by Sorensen. For their part, neither the Indian government nor the CBI had any firm tactical plan on how best to handle Nielsen's extradition. They fostered dispersed and isolated planning. R&AW, which did not run a station in the Indian Embassy in Copenhagen and had to rely on dispatching, from time to time, individual officers from its New Delhi headquarters to the Danish capital to gather 'intelligence', made little attempt to fuse its efforts with the CBI. The two agencies were unfocused and uncertain.

Stunned by the Eastern High Court's decision, the Indian government went on the offensive, obliquely criticizing the Danish judiciary's partisan rulings and then taking the ultimate step to freeze all diplomatic ties with Denmark. The *Indian Express* reported on 17 August 2011 that a decision had been taken to

> ...stall economic activities with Copenhagen and even decline meeting its officials following an advisory from Secretary (West) Vivek Katju in the Ministry of External Affairs. While the Ministry of Home Affairs has sent a circular to all sections to collate 'pending references with Denmark', the Finance Ministry has issued a communiqué on 'MEA's instruction relating to Danish projects' with the directive that all agreements and projects with that country be conducted only after 'prior consultations with the MEA'.[50]

A series of other measures, including non-issuance of visa to citizens of Scandinavian countries, NGO activists and officials wishing to travel to India for tourism or official business, followed. This was probably the first time in its diplomatic history that India, outraged and offended by the Danish Ministry of Justice's evasive action, acted tough, not mincing words to suspend hitherto cordial and friendly relations with Denmark. It was prepared to forego all diplomatic relations with a democratic country that was perceived as a harbourer of an international terrorist. Something was rotten in Copenhagen.

For Denmark, the enormity of India's strong and unequivocal step did not appear to have sunk in. It was prepared to protect an international terrorist on its soil at the cost of jeopardizing and endangering its relationship with India, with whom it had had long-standing economic ties. Was it simply a lack of political will in Denmark to extradite Nielsen or was pressure brought to bear on the judges and the Danish Ministry of Justice? What, or who, compelled the Danish Ministry of Justice not to challenge the Eastern High Court's decision in the Supreme Court? Was it a unilateral decision on the part of the Danish chief prosecutor? What prevented Sorensen from emphatically challenging the Eastern High Court's judgement in the Supreme Court? The Danes made no effort to explore whether the cabinet could have taken an executive decision to extradite Nielsen. The manner in which the Danish authorities, including the police and the judiciary, conducted themselves when they could have taken swift action against Nielsen (in detaining and handing him over to the CBI in accordance with the Interpol Red Corner Notice) when it was publicly known that he was living in Copenhagen, and their subsequent pussyfooting over the extradition issue, is ground for strong suspicion that the refusal to take timely police action against Nielsen was the result of pressure from a powerful Western state, which had good reasons to protect his real identity from becoming public. There appeared to be a pattern to the manner in which arms dealers and gunrunners are 'often protected from prosecution by their links to state intelligence agencies or other quasi-state actors'[51]. Did at least one Western power use its leverage over

Copenhagen to forestall Nielsen's capture and extradition so that the identity of the real end-users would forever remain a mystery? If so, then who really is Niels Christian Nielsen? Who was behind the 17 December 1995 arms drop over Purulia? Who were the weapons intended for?

10

A TRUE MANCHURIAN CANDIDATE

For most security and law enforcement agencies, intelligence gathering, even at its best, is an extremely confounding and perplexing business. Yet, there are individual cases where empty spaces in the minds of professional officers become repositories of bits and pieces of information; scraps and fragments which, for one reason or another, raise a slew of unanswered questions. At the beginning of 2000, Niels Christian Nielsen had surfaced in Denmark, but the criminal case against Peter Bleach and the Latvian crew members was virtually 'dead'. The Ananda Marga monks, indicted by the CBI for their complicity in the arms drop conspiracy, were still at large. The April 2004 arrest and subsequent interrogation of Tadbhavananda, alias Lalchand Parihar, secretary general of Proutist Universal and considered to be high up in the cult organization's hierarchy, did not yield any information on the intended end-users of the weapons. By the time Tadbhavananda died of cancer in November 2004, the CBI's interest in the arms drop case had waned. The subsequent transfer of the pugnacious Loknath Behera who, in his capacity as deputy inspector general, had achieved some breakthroughs, to the newly created counter-terrorism agency, the NIA, was an irreplaceable loss for the case.

Long before Behera's elevation to a senior position in the NIA, an intensive analysis by the intelligence agencies of the oddities and discrepancies in the long-drawn investigation had begun to throw up a plethora of unanswered questions. There were too many loose ends to ascribe the arms drop conspiracy to the Ananda Marga. Admittedly, some of the cult's monks were in league with Nielsen, but there was no intelligence on how and where they had escaped to. Unconfirmed reports indicated that some of the monks, including Satyendra Singh alias Satyanarayana Gowda (Randy), were given shelter by the Maoist Coordination Committee (MCC), an ultra-left extremist organization that had slowly begun to make its presence felt in southern Bihar and the belt which is now Jharkhand.[1] There were reports, too, that as the financial cost of MCC's protection mounted, Randy escaped to Nepal, but a team of CBI officers led by the investigating officer, DSP P.S. Mukhopadhyay, could not trace him either in Kathmandu or in Pokhara. As surely as the suspected link between the MCC and the Ananda Marga emerged, the theory that this particular Maoist outfit was to be the recipient of the Purulia weapons was debunked by the CBI as well as the IB. It was believed that while the MCC would readily accept AK-47 assault rifles and pistols, it would be 'impracticable' on the part of the outfit to procure rocket launchers and anti-tank grenades, which would be useless in their traditional fight against upper-class Hindu landlords in Bihar. 'Therefore, the possibility of these arms being meant for the MCC does not sound logical or probable,' the CBI concluded.[2]

To what extent and for how long the MCC provided shelter to the Ananda monks is still open to question, but it is almost certain that no left-wing insurgent group operating in Bihar or the gradually lengthening 'red corridor' at the time was the intended end-user of the weapons. The R&AW intelligence report of 25 November 1995 does refer to delivery of arms and ammunition to 'Communist rebels'. Bleach challenged this as a travesty of the information he had supplied to the British Special Branch, which had clearly reported to the Security Service that the 'arms were for use by anti-Communist insurgents against a Communist state government'[3]. While the R&AW's use of the expression 'Communist rebels' was, perhaps, the result of an oversight, Bleach's concern was that the 'report...that the arms were for Communist insurgents [was] a radical and fundamental change that could have had dire consequences had the arms in fact been successfully delivered'[4].

Following the brutal state repression in the late sixties and early seventies, the Naxalite movement led by the Communist Party of

India (Maoist-Leninist), which is the forebear of organized Maoist insurgent groups today, had been crushed. However, in the early nineties, the Naxalites in the region that is now Jharkhand had begun to quietly regroup and consolidate, though they did not possess any substantial arsenal. Most of the groups operating in Bihar were geographically limited to the Palamau and Chatra districts and had little influence in the southern Bihar districts of Dhanbad or Ranchi, leave alone the bordering Purulia and Birbhum districts of West Bengal. The expansion of the Maoist movement to states south of Bihar and subsequently Jharkhand, besides their significant capacity for military action and striking power, was a later phenomenon.[5] As late as 1999 and in the first five years of the twenty-first century, the Maoists relied on the limited strategy of raiding and looting police armouries and purchasing crude weapons manufactured in many of the clandestine production units in Uttar Pradesh, Bihar and Andhra Pradesh. The suggestion that the weapons were meant for the Maoist militants to enable them to destabilize the West Bengal government through large-scale violence in the run-up to the 1996 parliamentary elections does not hold water, especially because the CBI's investigation found nothing to link Nielsen to the Maoists. However, suspicion lingered that the Ananda Marga's channel would later have been activated to transfer the weapons to the left-wing rebels.

Another suggestion that the arms could have been meant for the Maoists in Nepal must be ruled out for the simple reason that left-wing radicals in the Hindu kingdom began their armed campaign against the political order with just two .303 Lee Enfield rifles in February 1996.[6] In fact, the Nepalese Maoists' fighting capabilities in the later years of the nineties was bolstered by looting police and security forces' armouries and weapons procured from illegal crude arms manufacturing units in India's Bihar and Uttar Pradesh.

The Bangladesh government outright dismissed any involvement of theirs in ordering the arms, and some journalists speculated on possible links to the Sri Lankan Liberation Tigers of Tamil Eelam (LTTE), but without any evidence. Utilizing an EUC that originated in Bangladesh was a safe cover, a ploy to give legitimacy to the arms deal and the safe transportation of the weapons by air from Bulgaria.

Under the circumstances, therefore, the best 'fit' as end-users was the Ananda Marga. The alleged involvement of some Ananda Marga monks and the drop zone being near Ananda Nagar appeared to dovetail neatly with this theory. Nielsen had indisputable links with the cult organization; he was in Purulia in the middle of September 1995 when he took photographs of the arms drop zone, including a two-storeyed building where some Ananda Marga monks were to await the parachutes to drop; and his 'confession' that he wanted to arm the 'poor, persecuted and tortured'[7] people (read Ananda Marga members) of the region were among some of the obvious clues. In its haste to prosecute Bleach, the five Latvians and Randy's brother Vinay Singh, the CBI singled out the Ananda Marga as the end-users who had 'spent approximately half-a-million dollars for the entire operation'[8]. As untenable as the Ananda Marga-as-end-users theory sounded, the CBI initially also suspected that the cult outfit organized the arms drop so it could 'sell' some of the weapons and ammunition to rebel outfits, including the Maoists and the insurgent groups in India's Northeast, to finance its foray into electoral politics in 1996 at a time when it was faced with a funds constraint. This was based on 'indications' that pointed to negotiations for one AK-47 rifle (missing from the Purulia consignment) in Bihar, where the price offered was Rs 65,000 or $1,600 at the prevailing black-market exchange rate.[9] This theory too was abandoned on the ground that the Ananda Marga would not have risked involving itself in the retail sale of AK-47 rifles, rocket launchers and rocket-propelled anti-tank grenades, landmines and night vision equipment to either the coal mafia in the Dhanbad belt or local criminals. Northeast India's battle-hardened insurgent groups (like the National Socialist Council of Nagaland [NSCN], United Liberation Front of Assam [ULFA] and the People's Liberation Army of Manipur), meanwhile, got their regular supply of sophisticated weapons from Kunming in China's Yunnan province till as late as 2004.

In fact, one of the biggest consignments of weapons 'destined' for the ULFA and the NSCN (Isak-Muivah), several times larger than the Purulia cache, was interdicted by Bangladesh authorities on 2 April 2004 in Chittagong. 'The consignment of new Chinese munitions was shipped from Hong Kong, then on to Singapore where more weapons, not Chinese-made, were added. The ship then continued to Sittwe on Burma's Arakan coast of the Bay of Bengal where the load was transferred to two smaller fishing trawlers, the *Kazaddan* and the *Amanat*, which ferried the weaponry to a jetty on the Karnaphuli River, Chittagong,' wrote Swedish journalist and Myanmar specialist Bertil Lintner.[10]

There were other obvious problems in branding the Ananda Marga as the intended recipient. If the sole purpose of the cult was to refill its shrinking coffers, it could have done so by simply appealing for donations from its large donor base in Western countries. On the other hand, the money Nielsen spent on purchasing the weapons and the AN-26 aircraft could have been deposited into the Ananda Marga's bank accounts abroad as well as in India, which would have given it solid financial footing to contest the 1996 parliamentary election. A former IB chief who, as a middle-ranking officer, had developed considerable expertise on the murky activities of the Ananda Marga, told me on a late November 2012 afternoon:

> If the Ananda Marga or its more radical wing, the Proutist Universal, nurtured electoral, and therefore, political ambitions at the time, it would have been naïve on its part to take on the Marxists in Bengal. I simply do not buy the theory that the weapons would have been sold to insurgent groups to fund the cult's electoral debut in the states of Uttar Pradesh, Bihar, West Bengal and Tripura.

However, the former IB director would not 'wish away' the Ananda Marga's capacity to commit violence, especially because it had a history of low-scale violence. Besides, he recalled that in the mid-nineties he had cautioned his headquarters of the gathering storm of left-wing extremism that had begun to appear in then undivided Bihar's Jehanabad, Aurangabad, Gaya, Chatra, Palamau, Dhanbad, Ranchi and Singbhum districts.

A former R&AW officer, who retired as a deputy chief of the agency in the later years of the last decade of the twenty-first century, is, however, convinced that the weapons were supplied by mercenaries for the Ananda Marga's political front, the Aamra Bangali, which wanted to strike back against the underground/terrorist organizations that operated in Assam and Tripura at that time. Other senior officers, who have since retired, described the theory that the Ananda Marga was the end-user as 'hogwash'. An Interpol officer, who refused to be identified, said emphatically that the weapons were definitely not for the Ananda Marga.

It is a fact that the Ananda Marga, despite being riven by factionalism, held sway over a nearly 10-square-kilometre swathe of territory, including its Ananda Nagar headquarters, in Purulia. Outside this sliver of land, it had to contend with the hostile CPI(M) that led the Left Front coalition in Bengal. A violent challenge to the prevailing political structure would have met with extreme levels of repression as well as tough legal measures, both by the Marxist state government and the Congress-led centre. It would have taken no time for the well-oiled, well-informed and efficient CPI(M) organizational machinery to obtain intelligence on the possible use of sophisticated weapons against it. The Ananda Marga could not have risked being clubbed alongside dreaded Islamist terrorist organizations and consequently invited a central government ban for the second time within twenty years of its existence. As a former senior IB officer, who would later assume charge of the R&AW, told me, 'The Ananda Marga could have done it back in the mid-seventies.'

More importantly, there were obvious contradictions in Nielsen's motivation to supply weapons to the SPLA and to the Ananda Marga, an organization with which he identified himself, at least outwardly. Nielsen has publicly stated that he supported the Ananda Marga for idealistic reasons. However, this does not seem to be very likely, given that he also had strong links with the largely Christian SPLA, which he has not acknowledged publicly in Denmark. It would be difficult to imagine that he would be part of two groups so diverse from each other from a cultural point of view.

According to Peter Bleach, Randy and his group did not organize the arms drop by themselves.

> Indeed, they have actually contributed very little, and what they did contribute was marked only by failure. According to

the original plan, the arms delivery was a complete success. No one detected that the AN-26 had deviated from its flight path, and if the group (read Ananda Marga) really controlled a large area around the drop zone, the inaccurate drop should not have mattered. The drop was well within the target area.[11]

Even if it is assumed that Nielsen and his Hong Kong-based Chinese 'business' partner, Wai Hong Mak, were the prime movers, it would seem unlikely that they were, in fact, the real people behind the scene.

By the time the AN-26 took off for its last flight from Phuket in Thailand on 21 December 1995, Nielsen had spent in excess of $600,000, virtually all of it in cash. Part of the money was allegedly provided by Mak, who was not an Ananda Marga follower and therefore would not have cared to provide large sums of money to purchase the arms for it unless he knew who the actual end-users were. Nearly two years after the arms drop, the Hong Kong police was able to provide Mak's profile and the extent of his involvement in the Purulia arms drop to the CBI. A resident of Helena Gardens, Prince Edward Road, Kowloon, Mak's Hong & Jen Trading Company sold gold bars, electrical appliances and cameras. He was once convicted in 1977 for obtaining pecuniary advantage by deception and in 1995 he was arrested for obtaining property by deception, but no charges were laid due to insufficient evidence. He was earlier suspected to be a member of the dreaded underground Chinese criminal network, Triad, but no such evidence was found. One of Mak's 'customers' introduced him to Nielsen, who subsequently purchased small quantities of gold bars from the Chinese on a 'number of occasions'.[12]

Sometime in September 1995, Nielsen suggested that Mak to join in partnership with him for transporting marine products to India, Pakistan and Bangladesh by plane. Nielsen also proposed that the two could use an aeroplane to smuggle gold into India. Mak told his Hong Kong police interrogators that he refused 'to take part in gold smuggling but agreed to join in partnership with Kim Davy in the transportation of marine products'[13], one of Nielsen's many cover stories sold to different people at different points in time ahead of the Purulia arms drop. The Nielsen–Mak partnership resulted in the purchase of the AN-26 in Riga, Latvia. Mak admitted that he transferred $200,000 (approximately HK$ 1,187,534, at the 1995 exchange rate) from the Hang Seng Bank (account no. 286-386511-001) to a Swiss bank whose account number was provided by Nielsen. He claimed he lost all contact with Nielsen after the plane was purchased in Riga, where he was present during the negotiations and final procurement. That is not true, because he did make an emergency visit to Phuket once the AN-26 had offloaded the weapons over Purulia to personally hand over $20,000 in cash to Nielsen. Contrary to Mak's denial[14] of being involved in the purchase of the arms and ammunition, his presence at the Bangkok meeting on 27 September 1995, when the drop zone was identified, was a clear indication of his awareness that a clandestine operation to deliver weapons to an Indian insurgent group was underway. For Interpol, Wai Hong Mak, was the 'most knowleadgeable [sic] about the individuals and interests behind the order of the arms delivery to Purulia'[15]. Interpol, which was apparently not satisfied with the manner in which the Hong Kong police authorities conducted Mak's interrogation, suggested that 'his motive to finance such an operation remains, however, unclear. What is the source of the half million dollars invested and were there any plans and terms for the reimbursement? The question of Mak's motive also raises hypothetical questions such as: was he acting on his own? If not, who and which interests were behind him?'[16] Hong Kong police do not have any information on the present whereabouts of Mak. What is known is that he sold off his Helena Gardens apartment to two Hong Kong-based Chinese in 2003.

The total amount spent for purchasing the weapons and the aircraft as well as other expenditure was a staggering sum, if it came from Nielsen's personal finances. To believe that, one must accept that he had great sympathy for the Ananda Marga. At the same time, it was known that he had close contacts with the SPLA and that he was involved in supplying weapons to the Sudanese rebels in exchange for gold resources in South Sudan. It is perhaps possible to believe that Nielsen had a lot of sympathy for one group of rebels within India, but to accept that he also held similar sympathies for a second widely different group is not realistic. This would suggest that Nielsen was acting as a front or a cut-out for an entirely different group or organization. Nielsen and some of his co-conspirators abroad may well have adopted outward signs of sympathy and conformity with the Ananda Marga to manipulate the cult for their own ends. By 2004, when the CBI prepared a fresh Nielsen dossier for the government, the agency appeared to have doubts about its earlier theory that the Ananda Marga was the intended end-user of the weapons, saying, 'There are many missing links in the international conspiracy. We also have to find out the real end-users of [the] arms and ammunition.'[17]

The Sharjah Missing Link

Among the inert mass of secret documents and confidential files was a huge loose end that neither the CBI nor the intelligence agencies looked into thoroughly, though it begged urgent attention and investigation. Even as the clue, no matter how distant or seemingly innocuous, offered itself, there was lethargy and a lack of will to follow up on it. This was Nielsen's sudden decision to visit Sharjah in the United Arab Emirates (UAE) immediately after his return to Burgas, Bulgaria, from the Varanasi 'dry run' towards the end of November 1995. The Latvian pilot, Alexander Klichine, flew the AN-26 from Karachi to Sharjah sometime between 7 and 9 December 1995.[18] Nielsen flew to Sharjah for he had immediate 'business' to attend to there. Several witnesses[19] in Latvia have revealed that after the AN-26 deal was struck with Latavio, Nielsen purchased a large quantity of aircraft spare parts which were flown to Plovdiv and then disgorged in Sharjah. But there is no clue whom the aircraft components were intended for. At the desert sheikhdom, Nielsen and the crew put up at Carlton Hotel. However, it is most likely that Nielsen decided in Sharjah that the weapons-laden aircraft would take-off from Burgas on 10 December 1995, and accordingly instructed Peter Bleach to advise the UK-based Baseops Europe to draw up a flight plan.

'The aircraft was definitely in Sharjah because Davy telephoned me from there and gave me a proposed route to Bourgas to file with Baseops,' Bleach wrote in his statement to the CBI.[20] Who in Sharjah did Nielsen consult with? Did he have associates in Sharjah who were familiar with the air route that clandestine flights usually took to deliver illegal shipments of weapons elsewhere in the world? The only effective network of arms dealers which had some basic and operational presence in Sharjah was that of Viktor Anatoliyevich Bout, the Russian gunrunner who would later rise to infamy as the 'Merchant of Death'[21] for the mammoth scale and extent of his operations in arming rebel groups in the African countries of Angola, Sudan, Liberia, Sierra Leone, Democratic Republic of Congo, and Togo, besides the Taliban in Afghanistan, Muammar Gaddafi of Libya and the Fuerzas Armadas Revolucionarias de Colombia (FARC) or The Revolutionary Armed Forces of Colombia.

There is evidence to suggest that Bout, an alleged former KGB (Komitet Gosudarstvennoy Bezopasnosti; the Russian state security apparatus) major who went with at least five aliases and several passports, established his 'business' in Sharjah in 1993, although he ran parallel operations from Ostend in Belgium between 1995 and 1997.[22] But for the massive scale of Bout's operations, his and Nielsen's modus operandi was near identical. Bout would sign contracts with weapons suppliers in Western European countries, Israel and a few firms in African states. He would supply them with forged EUCs obtained fraudulently from small African countries like Togo and Burkina Faso. The EUCs would then be dispatched by email to the Bulgarian KAS Engineering Co. (involved in organizing the Purulia weapons shipment in Bulgaria; the company later shifted base to the offshore haven of Gilbraltar) which would then sub-contract the deals to Bulgarian arms manufacturing companies, including the Kazanlak-based Arsenal Ltd and VMZ, or Vazov Machine-Tool Producing Plant, based in the town of Sopot, among others. The weapons manufacturing factories would label the shipments 'technical equipment' (as the boxes containing the Purulia consignment were marked). Russian Antonov or Illyushin and other cargo aircraft that Bout operated would invariably be loaded with the weapons shipments at Burgas, from where the majority of the flights would head directly for east and central Africa, transiting via Nairobi, Khartoum and Johannesburg.[23] On one occasion, East European Shipping

Corporation, a firm based in Nassau, in the Bahamas, was the broker of a weapons shipment that was transported by ship to Lome, Togo. East European Shipping Corporation was represented in Europe by a company known as Trade Investment International Ltd[24], located in the UK and Northern Ireland, one of three companies owned by Samuel Sieve (formerly known as Shlomo Silberstein), the London-based arms dealer of Polish origin who Peter Bleach put in touch with Nielsen.[25] As far as his alleged link with Nielsen and the Purulia arms consignment was concerned, Samuel Sieve, of Cumberland Mansions, George Street, London W1H 5TE, was never questioned by the British authorities.

Just as Nielsen supplied arms to Sudan in exchange for gold rights in the southern part of the country, Bout provided weapons to the anti-Communist UNITA in exchange for diamonds mined in the Cuango Valley.[26]

There was yet another operational similarity between Nielsen and Bout. Both had set their sights to expand their operations in South Asia, with a base in Dhaka, Bangladesh. While Nielsen failed to realize the plan to enter into a business partnership with a Bangladeshi firm called Air Parabat Limited[27] for the proposed joint venture company called Cargo Air Transport Services, UN investigators who probed Viktor Bout found that in May 2000, five years after the Bangladesh airport authorities refused landing permit to the suspicious AN-26 aircraft, the civil aviation authority of that country granted one Omni Airline an air traffic operator's licence to operate as an international cargo airline pursuant to the government's liberalisation of the aviation sector. The newly formed airline was a joint venture between the Enem-Omni companies of Bangladesh and Viktor Bout's UAE-based Air Cess cargo airline. It planned to operate three weekly flights to Sharjah, Delhi, Mumbai, Chennai, Kolkata, Kathmandu, Hong Kong, Bangkok and Singapore with 15-35-ton capacity freighter aircraft.[28] Clearly, by the time he was arrested in a sting operation codenamed 'Operation Relentless' by undercover special agents of the US DEA in a Bangkok hotel in March 2008, Bout, five years younger to Nielsen, had built an empire. Back in the early and mid-nineties, Nielsen was small fry compared to Bout's large and truly global gunrunning enterprise.

Yet another loose end that the CBI and Interpol did not adequately pursue was the role played by Border Technology and Innovations UK Ltd, an arms dealing company based in Abingdon, Oxfordshire, which Peter Bleach had contacted to make the final supply of the weapons from Bulgaria. Managed by two former British military officers, Peter Scott and Robin Campbell, BTI had supplied weapons and military hardware in Angola in 1992, though it is not known whether Viktor Bout had ever contracted the firm for supplying arms and ammunition to the UNITA. BTI claimed to have appointed agents in twenty-two countries since 1990 and had secured contracts with military and civil defence units in at least fifteen countries, including Bangladesh, Sri Lanka, Pakistan, Lebanon and Venezuela.[29] The British police did question Scott and Campbell, who simply provided factual information specifically related to its dealings with KAS Engineering Co. and the Bangladesh EUC, and the payments it received by banker's drafts once the Purulia consignment was loaded onto the AN-26 at Burgas.

Of course, the uncanny operational similarities that I have cited could be dismissed as too tenuous to be relied upon, but the fact remains that Nielsen's Sharjah link and his meetings, whoever they might have been with, was papered over or just got buried in the mountain of documents that the CBI had obtained as a result of the execution of the Letters Rogatory by several countries to which the footprints of the Purulia arms deal were traced. For a few years from the first days after the 17 December 1995 arms drop, the CBI's probe and the R&AW's intelligence gathering was reflective of episodic bursts of meticulous investigative work matched by their prolonged indecisiveness to pursue leads abroad and an utter lack of a sense of history.

Viktor Bout's CIA Links

Flying cargo aeroplanes laden with lethal weapons and violating international and domestic laws requires more than individual effort. It takes an internationally organized network of individuals, well funded, well connected and well versed in brokering and logistics, with the ability to move illicit cargo around continents without raising the suspicions of law enforcement and regulatory agencies across countries or with the ability to deal with unforeseen obstacles.[30] Bout's arrest by the Thai police, acting in concert with US DEA special agents, in March 2008, was preceded by extensive investigations by the UN that unravelled the mystery of the arms shipments to several African countries against which sanctions had been imposed for the import of military hardware that potentially fuelled the bloody civil wars there. It was discovered that at the centre of these investigations were some East European and Russian arms dealers, the most prominent being Viktor Bout 'who also worked for the Ameircans before running foul of them'[31]. From the time the UN Security Council received the investigation reports until 2008, Bout was under the scanner of American investigators who were intrigued by his professional background as a former KGB major and his web of UAE-based air freight companies that had supplied the Taliban with massive quantities of weapons in the mid-nineties. After Bout was extradited to the US under controversial circumstances—Thai authorities were under tremendous pressure to do so—several analysts observed that his 'testimony could prove highly embarrassing to his captors, who, it can be assumed, will attempt to prevent him speaking about his work for a variety of US government agencies'[32].

By the late nineties, acting under US pressure, several countries in Europe and Africa, where Bout had business interests, started criminal proceedings against him, closing some of his front companies and forcing him to flee; first Ostend in Belgium, then Sharjah and finally Johannesburg, before he found safe sanctuary in Moscow in 2003. Even as the UN acknowledged, in the context of Angola, that international trafficking in arms and ammunition required more than individual effort, there is not a shred of paper to suggest that the experts' panels that investigated the sanctions-busting weapons deliveries by Bout and his accomplices to organized insurgent groups in Africa ever took into account the Purulia arms drop case. Nor did they consider the means and methods applied by Nielsen to procure the weapons from Bulgaria, the purchase of the AN-26 aircraft in Latvia, the hidden structure of his network sheltered behind layers of shell companies and, most importantly, the Dane's clandestine arming of SPLA rebels in exchange for gold and gemstone mining rights in South Sudan, stretching to bordering Kenya and Uganda. Although the Purulia arms drop case was one of the first well-known instances of international gunrunning, by 2005 it had become either a one-off forgotten episode, even among member states of Interpol, or there was something more sinister to it for which powerful countries, including the US and UK, displayed reticence in pursuing Nielsen, who was wanted across five continents.

MI5's Silence

'While the Purulia arms drop case was a real challenge, we did everything we could and we did manage to dig up quite a bit. We looked into all possible angles, including the involvement of foreign intelligence agencies,' recalled a former R&AW chief who, as a joint secretary in the Indian external intelligence agency at the time, was intimately involved in the manhunt for Nielsen. After executing the Letters Rogatory, the British shared no further information beyond that which the CBI had already requested for or even on the basis of the liaison relationship with the R&AW. Indian government files suggest that two other British nationals, Benson Patrick and Jeremy Weatherhill, who ran a private security company called Lynx Security Services at Cowcross Street in London, were 'known to Peter Bleach' and were 'aware of all details of the Purulia arms drop'.[33] There is nothing to suggest that any effort was made by either the CBI or Indian intelligence to find out the antecedents of Patrick and Weatherhill and to what extent they knew about the arms drop. The R&AW officer who was in charge of liaison with the Security Service SLO in the late nineties would regularly report to his superiors at the agency's headquarters that his source had dried up. The US Department of State's Protective Intelligence Division had furnished the CBI with Nielsen's fingerprints that their agents had been able to lift from a Customs declaration form the Dane had submitted at Newark airport in New Jersey during his visit in July 1995. Details of two federal crimes that Nielsen had committed in the US—passport forgery and printing counterfeit $100 bills—were also supplied to the CBI. On its part, the R&AW sought the cooperation of the CIA, which offered little assistance, even though the two security organizations were tied by informal liaison agreements to share intelligence. 'Yes, the CIA was approached too,' the former R&AW chief admitted, but was unwilling to share the details. His colleague Jayant Umranikar, the senior R&AW officer stationed in Mumbai when the arms drop occurred, was more forthcoming. 'The CIA knew about it, but at the time the Americans did not provide any information or assistance.'[34] In any event, the CIA's suspected role came under intense scrutiny as part of the global investigations by the R&AW–CBI–Interpol.

At the same time, the British silence was puzzling, especially after the proactive role the MI5 had played in debriefing Peter Bleach. As described earlier, between August and October 1995, playing the role of an astute double agent, Bleach had kept the police, and through the Special Branch, the MI5, informed of the details of the conspiracy and the deal-in-progress. MI5, which by this time had evolved more into a counter-intelligence and counter-terrorism organization in the wake of the dissipation of the Soviet threat, played its part by sharing intelligence on the Purulia arms drop conspiracy through its long-standing security liaison office at the British High Commission in New Delhi with the R&AW. Intelligence, deemed to be high grade, was shared on 10 November, 17 November and 15 December 1995. For some strange reason, the MI5 deliberately did not share the intelligence with the R&AW London station chief (a joint secretary-level officer).

True to its trade, MI5, acting through the Special Branch and the North Yorkshire police, extracted almost all details—the quantity of weapons and ammunition, the volume of money that was involved in the transaction, the names of the bank from which money was transferred to facilitate the purchase of the weapons, the names of the principles, including Kim Palgrave/Peter Davy, the unnamed insurgent group to which the weapons were to delivered, the exact coordinates and grid reference of the location and Bleach's concern that 'he may be being set up by someone possible [sic] the American[s]'[35]. The two British policemen, Detective Sergeant Stephen Leslie Elcock and Detective Constable Jones, outright described Bleach's suspicion of some American involvement as 'outrageous', but we do not know whether MI5 would have shared this 'outrageous' and unconfirmed piece of information with the R&AW. R&AW officers who served at the time and supervised some of the intelligence operations in several countries to supplement the CBI's investigations later told me it would have been highly unlikely that the British security agencies would not have shared Bleach's information and his very specific concern with their more powerful and resourceful cousin, the CIA.

Bleach appeared to have raised several questions in a letter addressed to the UK Home Office, which he wrote from his cell in Calcutta's Presidency Jail on 23 January 1998. A 10 June 1998 letter from G.W. MacAlister, an official of the Home Office Organised and International Crime Directorate, to the then British deputy high commissioner in Calcutta, Simon Scadden, made a reference to these questions, but invoked the plea that some of the queries fell 'into areas on which the Government does not comment'[36]. While Bleach's information on the ongoing arms deal was duly passed by the Security Service, 'acting as a focal point' to 'interested government departments', presumably its sister organization, the MI6, or the Secret Intelligence Service (SIS), it was shared with not just the Indian R&AW but also with the PET, the Danish Security and Intelligence Service, since four Danish nationals were suspected to have been involved when the plan to procure the arms was at its nascent stage. At that time, PET operatives tracked down Haestrup and reportedly dissuaded him from going ahead with the plan to procure the weapons. The Dane did not pay any heed; the deal went ahead. There is nothing on record to suggest that the PET shared the information with the R&AW or coordinated its effort with the Indian intelligence agency to prevent the arms drop plan from going underway. Danish intelligence showed no interest in tracking Nielsen, even though he was moving in and out of Denmark quite freely in the months preceding the arms drop.

The heavily redacted letter of MacAlister to Scadden makes it abundantly clear that the R&AW were informed about the Purulia plan following the North Yorkshire police's meeting with Bleach on 14 September. 'After background enquiries had been completed, on November 10, 1995, the Indian authorities were informed as to the possibility of a delivery of arms and ammunition to the Dhanbad area. Further reports were sent to them on November 17 and December 15, 1995.'[37] And then MacAlister's letter made an interesting revelation. It said, 'At no stage did the Indian authorities request that enquiries into this matter should be discontinued,'[38] indicating that the British Security Service pursued the matter and possibly learnt in greater detail the nature of the delivery (whether it was for a terrorist or insurgent group), the background of the conspirators, their capabilities and perhaps even the intended recipients of the weapons. If it did succeed in finding out who Nielsen actually was and for whom the arms were meant, the Security Service did not share the information with the R&AW. What can, however, be said with certainty is that MI5 knew a great deal more in the months after the arms deal than it knew as a result of Bleach's cooperation. 'An R&AW officer who liaised with the Security Service would report that his source had simply shut up,' a retired senior R&AW officer told me in the course of a lengthy interview in November 2012.

Interpol's attempts to elicit information from the British authorities turned futile. What was it that the British suddenly wanted to conceal after it had willingly shared substantial information with the R&AW in November and early December 1995? The MI5 had, through the North Yorkshire police, told Bleach to withdraw himself from the venture, though Bleach did not do so. He continued to 'run' with the deal. Why did the MI5 or the British police not act to prevent the arms drop, especially when part of the conspiracy had taken place on British soil—London—where Nielsen and Bleach had signed the contract for the purchase of the weapons on 13 September 1995? The Interpol had identified an Irishman, 'Mick', as Leonard Paul Francis, alias Perry Mick Joseph, born on 9 December 1956, in Dublin. It was Mick who met the BTI representative, Robin Peter Campbell, at St Giles Hotel on Bedford Street in London on 19 December 1995, to hand over to him two bankers' drafts ($75,000) drawn on HSBC. Since Mick, Nielsen's partner-in-crime in Norway, was under instructions not to release the bank drafts without authorization, Campbell waited till evening when he met Mick in his hotel room, where Nielsen called up to hand over the drafts to the

BTI representative.[39] Months later, an email from an Interpol officer to a counterpart in India revealed that British intelligence had for long been 'keeping an eye' on Mick, who was suspected to be linked to the IRA.

The R&AW?

Nielsen's sensational 'disclosure' in April 2011, that the arms drop operation was a joint R&AW–MI5 effort to topple the Jyoti Basu-led Left Front government by instigating violence, which could then be used as a pretext by the Congress government of Prime Minister P. V. Narasimha Rao to impose president's rule in the state, was fantastic, if not total hogwash. If it was indeed an R&AW operation, Nielsen could have named his handler in the Indian intelligence agency. He could have revealed when he was recruited as an agent, who did he report to, what were the channels of communication with his handler, how many times did he meet his case officer, who supplied the huge amounts of money to purchase the weapons and the aircraft and how did the money reach two Swiss bank accounts, besides identifying his associates in Europe and the US who assisted in the arms drop. The Times Now editor-in-chief, Arnab Goswami, who interviewed him at length, did not bother to ask him these very basic questions, allowing Nielsen to spin his web of lies and deceit aimed at creating an appalling stink that had some media purchase. Regardless of his allegations, it was perhaps not easy for Nielsen to wear the mask and at the same time point the finger. Nevertheless, because of his 'revelations', the Purulia affair briefly flickered back into life in the summer of 2011.

Safely ensconced in his home country and protected by the Danish authorities once the courts there had ruled he would not be extradited to India, Nielsen's 'disclosure' on Indian television was an orchestrated misinformation campaign. He was having fun at the expense of the R&AW, the CBI and the Indian government in general, thumbing his nose at them, mocking at their impotency and inability to catch him. His allegation that Pappu Yadav, MP from Bihar, who had a criminal background, helped him reach the Indo–Nepal border, had some grain of truth. In January 2002, a US official tipped off an Interpol officer in Europe, indicating Pappu Yadav's alleged involvement in assisting Nielsen. The US official had extracted this information after tracking down and questioning Robert Dean Child (in the US). Requesting the European Interpol officer to treat the information as 'highly classified,' the US official was believed to have said that Child disclosed Pappu Yadav's role in helping Nielsen escape after asking, 'How much trouble am I in?' I have had access to a January 2002 email that named Pappu Yadav, but my source was unwilling to share any other details of the 'highly classified' information that he had received from the US official. A source in the CBI who investigated the arms drop case for years emphatically denied Pappu Yadav's role, dismissing my query on his likely involvement with a sharp and firm 'No'. It could also be that Nielsen may have shared Pappu Yadav's name with Child since he had heard it earlier from Randy. Officially, the CBI as well as the Indian government denied the hand of any political leader in the arms drop, saying, 'His [Nielsen's] self-serving allegations and attempt to give a political colour to his crime and thus deflect the judicial process of his extradition is not substantiated by the evidence and facts.'[40] However, R&AW sources have revealed to me Pappu Yadav's alleged role in supplying small arms to the Nepalese Maoists in the nineties. When I contacted Pappu Yadav's wife, Ranjita Ranjan, over phone, she said she 'did not want to comment as we are not interested in the topic'[41] before hanging up.

Nielsen's claim that Pappu Yadav, who had—and has—a criminal background,[42] had worked in collaboration with the Prime Minister's Office to bring in the arms so that violence and chaos could be engineered in West Bengal as a prelude to imposition of president's rule, was indeed 'fanciful'[43]. A pariah politician from the badlands of Bihar's Purnea district, Pappu Yadav simply did not enjoy any clout with Narasimha Rao's PMO. Inquiries by the CBI had indicated that Yadav and his family, including his father, had, and may continue to have, sympathies for the Ananda Marga, but no more. Nielsen's allegation on Indian television in April 2011 that the then Prime Minister P.V. Narasimha Rao approved the plans to use the R&AW to bring in weapons so that an insurgency could be precipitated in West Bengal, which would then give the central government the opportunity to directly rule over the state, does not quite fit in with his claim in his book that he wanted to arm 'the poor, persecuted and tortured local residents' (of Purulia) against the 'brutal militia forces' of the communist regime.[44] He could not have acted then both as an R&AW agent employed to drop the weapons over Purulia so they could be used to meet Narasimha Rao's alleged goal of engineering violence and chaos and at the same time as an Ananda Margi idealist out to supply weapons to the people of Purulia so they could defend themselves from communist violence. More importantly, Narasimha Rao's government enjoyed a comfortable relationship with the CPI(M), which provided considerable political support to his regime at the centre.

Jayant Umranikar, the R&AW's special commissioner in Mumbai when the arms drop took place, and who headed the agency's Mumbai end of the investigation after Nielsen's escape from Santa Cruz airport, emphatically denied that the Indian external intelligence organization was 'in any way involved' in the weapons drop.[45] 'For sure it was not an R&AW operation,' Bahukutumbi Raman, who was in charge of the agency's high-grade special operations as well as liaisoning with foreign intelligence agencies before he retired in early 1995, asserted.[46] So did my father, Bibhuti Bhushan Nandy, who was an additional secretary in the R&AW and, by virtue of that position, would have had a fair idea whether the agency he worked for was at all involved in the Purulia arms drop.

Yet, it must be granted that in the heavily compartmentalized R&AW where, by careful design, officers up and down the hierarchy knew little about one another's operational or analytical work, neither Umranikar nor Nandy, nor even others could have been definitively sure whether the top management had any involvement in the arms drop. Whatever Nielsen's ploy was worth, the fact is that the R&AW was, at least at that time, quite incapable of organizing and implementing such daring operations spanning across five continents.

There was indeed a time, way back in the late eighties, when individual operatives carried out operations in India's neighbourhood, especially in Myanmar where helicopters taking off from Assam's Doom Dooma air base of the Aviation Research Centre (ARC), the

R&AW's aviation and air surveillance wing, would land in areas under the control of the Kachin Independence Army (KIA) insurgents, to supply arms procured from Southeast Asia. On a few occasions in the early nineties, the R&AW supplied large consignments of weapons, non-lethal military equipment and huge amounts of cash to the Northen Alliance in Afghanistan led by the ethnic Tajik, the 'Lion of Panjshir', Ahmad Shah Massoud. Much earlier, in the fifties and sixties, Indian intelligence (the R&AW was yet to be created) would offer logistics support and air transport to ferry weapons supplied by the CIA to certain Tibetan tribes fighting the Chinese authorities.[47] Helicopter flights, jointly manned by the CIA and Indian intelligence, would take-off from Kalimpong in West Bengal's Darjeeling district to airdrop arms for the Khampa resistance fighters in Tibet.[48]

But to say that the R&AW planned the Purulia arms drop in collaboration with the MI5 (primarily a counter-espionage and counter-terrorism agency, the Security Service's charter does not allow it to launch intelligence operations abroad, which is the domain of the MI6), would be to give too much credit to India's foreign intelligence agency, which was institutionally averse to taking risks, a factor that had atrophied its capacity to undertake covert action. Across different parts of the world, wherever it has stations, R&AW operatives work under diplomatic cover. In the early nineties, barring a few stations such as Dhaka, Kathmandu and Islamabad, the R&AW was not known to use 'illegals' or agents who did not enjoy the benefit of operating under diplomatic or official cover. As an intelligence organization, it had and continues to have access to unaccounted-for secret funds to finance its operations, most of which go unused. Whatever is used goes into agent running, pay-offs to agents and sources with the aim of subverting the loyalties of citizens of other countries to procure classified documents, and plan and execute small-time operations mostly in Pakistan (especially in Balochistan), Bangladesh, Nepal and Thailand. Special operations of the Purulia kind would be unthinkable even among the R&AW's top brass, for covert activities of such scale and dimension often needed approval at the highest political level. When a senior R&AW officer, after assessing his capabilities, proposed a 'hit' on Dawood Ibrahim in Dubai, where the Mumbai underworld don would often visit from his ISI-sheltered Karachi sanctuary, the plan failed to get clearance at the highest political level. On another occasion, then Prime Minister P.V. Narasimha Rao refused to give approval to a 1996 R&AW plan to airlift to India the beleaguered and defeated communist president of Afghanistan, Mohammad Najibullah, who was subsequently tortured and killed before his body was strung up on a lamp post in central Kabul by the marauding Taliban forces. By the late nineties, the R&AW, as India's intelligence agency, did not attract extraordinarily talented and motivated officers capable of undertaking covert missions in India's near-abroad or other parts of the world to serve the country's foreign policy objectives.

If the R&AW was indeed behind the arms drop, there was no reason for it to have shared intelligence on it with the IB and the home ministry twenty days in advance. Besides, the post-arms drop intelligence and information that it supplied to the CBI ran into several hundred pages. The result of some of the inquiries it made in a number of countries was accurate, though some were off the mark, as it almost always is in cases involving no less than fourteen countries, some of them non-cooperative, if not outright hostile, towards Indian interests. The R&AW could have avoided a lot of trouble by procuring the weapons from registered and government-authorized Bulgarian arms manufacturers. Kalashnikov rifles, matching ammunition, rocket-propelled grenades and landmines were cheaper and readily available in some of the Southeast Asian countries, including Thailand and Cambodia, from where a bulk of the supplies were purchased for delivery to the KIA rebels in the jungles of the upper reaches of Myanmar bordering China. Clandestine shipments of weapons from the flourishing arms bazaar in Southeast Asia would have helped the R&AW achieve the twin objectives of cutting costs as well as distance, besides ensuring reliable and assured delivery to the end-users.

Bleach's CIA Hypothesis

While the CBI followed a compartmentalized approach towards its investigations, sticking to collecting evidence that could be produced in court to further its aim of prosecuting those who had been arrested—Bleach and the Latvian crew members—it also enlisted the R&AW's expertise in unearthing the mystery. The CBI also fell back on Peter Bleach for insight into who Nielsen really was and who could be behind the arms drop. In his statement to the CBI, Bleach confidently claimed that Nielsen and his associates 'are very careful to work through other people to achieve their aims and to conceal their identities'. This, according to Bleach, 'would suggest a foreign government or intelligence agency... The points above would seem to remove most Western European governments from suspicion... Russia would seem to be preoccupied with its own troubles and is not these days noted for throwing large sums of money around. China is an unknown entity from my point of view as are most other countries in the region'. According to Bleach's analysis concluded that 'this leaves Pakistan and America'. He 'could not help thinking that if Pakistan wanted to do this, it could have done it much better all by itself... Altogether, America does seem to be the likely candidate. From the beginning, I thought that the entire project had an American "feel" about it... Davy's residence in America does tend to support some kind of American involvement. Whether it is official, unofficial or non-governmental involvement is an entirely different matter.'[49] Seventeen years later, Bleach, sixty-two years of age, still 'believes' that the arms drop was an American operation, though, 'I have no documentary evidence to prove it.'[50]

Bleach's rudimentary and simplistic hypothesis was an eye-opener for the CBI, especially after information unearthed by Indian intelligence indicated that Nielsen visited the US four times before the arms drop. He was able to travel unhindered each time, despite being wanted in that country for two federal crimes. On 30 July 1994, he entered the US, landing at New York's JFK Airport, and left via San Francisco on 5 June 1994. His next visit to the US was noticed to be on 20 April 1995, when he landed at San Francisco and proceeded to Colorado. On this occasion, there is no record of 'Kim Davy's' exit from the US. On 18 July, Nielsen landed at Newark airport, New Jersey, and was in New York the next day before reaching Los Angeles on 26 July 1995. On 6 August, Nielsen entered the US at the airport

in Cincinnati and spent a few days in Colorado before leaving via
Minnesota on 15 August 1995. His suspected presence in Colorado in
1998, which was attested to by the then Las Animas county sheriff's
deputy who had identified him from a photograph, reinforces the
suspicion that even though he was wanted in the US, no effort was
made to detain or arrest him. Nielsen's ranch in Colorado, as Interpol
subsequently discovered to its shock and horror, operated as a deep-
cover, clandestine paramilitary training facility which he headed.
Recalling the investigation reports that his deputy had filed in 1998,
the former sheriff of Las Animas, Lou Girodo, said,

> This ranch, which had some military training camps, was
> located at a place called Gulnare, west of Aguilar town and
> just off the Highway of Legends (State Highway 12). The
> deputy's reports indicated that the group possessed some
> automatic weapons. When we tried to investigate further,
> the men disappeared rapidly. Bill Loose moved out to another
> county in Colorado.[51]

From its initial reluctance to venture into the world of spies and
spooks, the CBI veered around to giving it some credence as it found
layers and layers of unique events, players and sub-plots resembling a
carefully crafted intelligence operation. What the CBI did not know
was that the IB had maintained discreet files since the early seventies
on the suspected links between the CIA and the Ananda Marga, of
which Nielsen was a member. However, parallel to its official
investigation of a terrorist offensive, at an 'unofficial and discreet'
level, the CBI sought the assistance of the R&AW, which was only
too glad to take on an inquiry of a perceived Western intelligence
operation and, if possible, to identify and unmask the plotters and the
intended weapons' end-users. Curiously, every piece of intelligence
the R&AW stations generated and the agency's headquarters shared
in writing with the CBI was marked in the subject field as 'Attempts
by an Indian insurgent group to acquire arms and ammunition'.
Nevertheless, what the R&AW and Interpol dug out, separately, in
their exhaustive inquiries between 1995 and 2000, was unexpected
and therefore startling. 'At first, the information the British supplied

to us read like a gravy plot. But once the arms drop took place it was serious business. There was a steady flow of intelligence. Hundreds of leads and names cropped up. It was like peeling an onion,' recalled an R&AW officer, who has since retired from service. The gush of intelligence between 1996 and 1998, often at regular intervals, indicated that a host of American and British nationals, using a web of aliases and fake passports, were involved with Nielsen and the arms drop.

Nearly a year after the arms drop and Nielsen's escape, it came to the notice of the R&AW that an American passport holder, Robert Newman, who identified himself as a financial manager, had in 1993 taken over as a director of the Hong Kong-based Howerstoke Trading Ltd, the company with which Nielsen was associated and which was suspected to be involved in gold and gemstone mining operations in South Sudan. Newman, who was born on 22 June 1955, had once been arrested in New Delhi in March 1990 in a smuggling case, when the passport he used at the time was in the name of Robert Amber, with whom Nielsen was in constant touch over email. The R&AW subsequently found out that Newman was issued a one-year business visa by the Indian mission in Hong Kong on 28 July 1995. After his arrival in Calcutta, Newman, whose permanent address recorded in the visa application form was Sarasota, Florida, checked in as a guest at Hotel Astoria on 17 November 1995. Another person, who held a British passport (no. 017180312) had also checked into the same hotel along with Newman, but did not enter his name in the guest register. A thorough check of all visa applications made by Europeans and Americans at the Indian High Commission in Hong Kong threw up the name of one Andrew Wentworth Ping who visited India on a six-month tourist visa issued to him on 9 November 1995 by the mission.

On 17 November 1995, presumably the day the two checked into Hotel Astoria, Robert Newman received a telephone call in Room No. 8 from a Hong Kong number (852-2787-6347) belonging to World Trading Company, linked to Wai Hong Mak. As far as the R&AW was concerned, the presence of 'two close associates' of Nielsen in Calcutta exactly a month before the Purulia arms drop, and the fact that their telephone number in a Calcutta hotel was contacted by Wai Hong Mak, was 'significant'. Subsequent investigations revealed the presence of Newman and Ping in India when the arms drop actually took place. This left the R&AW wondering which of the two men acted as Nielsen's handler. In his visa application form, Ping claimed to have earlier visited India in April–May 1995, but was keen to visit Delhi, Agra and Jaipur during his November 1995 trip. Likewise, Newman claimed that during his previous trips to India between December 1994 and January 1995, he visited Delhi, Mumbai, Calcutta, Allahabad and Varanasi.

Ping, like Newman, was also 'closely associated' with Nielsen, and had two other British passports, one issued in London on 7 April 1993, and the other by the British Embassy in Bangkok on 19 November 1993. In his visa application form, Ping recorded his permanent address as Pearl City Mansion, 33, Patterson Street, Causeway Bay, Hong Kong, the same address as Nielsen's. On the other hand, Newman was traced to two US addresses—one in Sarasota, Florida, and the other at 7627, 16th Street, NW, Washington, DC. His links with Nielsen were suspected to be even stronger when it was discovered that at the time of the Purulia arms drop his 'present' address was Belomorskaya Road 27, Apartment 45, Khabarovsk, the heavily militarized Russian city in the Far East, where Nielsen lived for two years towards the end of the Cold War. But what intrigued the R&AW was the identical dates of birth of Newman and a man by the name of Peter Brook (22 November 1955), who was found to be the principal holder of a Citibank Visa card (no. 4568-8170-2002-7515) that expired in October 1996, which Nielsen used as a supplementary card. Brook held an Australian passport (no. 2081849), though his address was in Raffles City, Singapore, which was also the address entered into the records of a UK-based company, Leahurst Limited, whose sole director was Andrew Wentworth Ping. The company, which was established in September 1991, was dissolved before the Purulia arms drop.

The more the R&AW tried to break through the intricate maze of fictitious names, shell companies and common locations, the more it got sucked into the stratagems of the Purulia plotters. In the course of identifying the numerous telephone calls that Nielsen made to his associates in Hong Kong, different cities in India, Pakistan, Singapore, the UK, Denmark and a few South American countries, the R&AW chanced upon a vital clue that seemed to confirm their suspicions that the arms drop was more than the work of a few international smugglers or Ananda Marga sympathizers. A link to the Vessel Satcom System was located in Nielsen's Hong Kong flat. The instrument, hired from a British company, was the size of a suitcase, which could be physically carried and placed either on a sea-faring vessel or even on land, to make long-distance telephone calls as well as other data transmission. A number of phone calls were made using the Vessel Satcom (number 682-341-927) to Nielsen's room in Pearl Continental Hotel, Karachi, between 13 and 17 December 1995, when he, the five-member Latvian crew (excluding the two Russian ground engineers), Bleach and Deepak Manikan lived there prior to flying out for the arms drop mission. It was suspected that the Vessel Satcom System had an attached encryption device which prevented any data interception.

The discovery of the Vessel Satcom System link in Nielsen's apartment and its operational use provided an extraordinary and deep insight into the network that was suspected to be behind the arms drop. Calls made to Nielsen using the Vessel Satcom System between 15 November and 18 December 1995 were traced. Towards the end of February 1996, the System was located somewhere in Asia and analysis of the phone numbers contacted from the System between 12 and 17 December 1995 indicated that it was neither with Nielsen nor in his Hong Kong apartment. It was not found to have been linked to any sea-faring craft and was used to make calls via IMMARSAT[52].

While Howerstoke Trading Ltd was clearly the cover company, the men associated with it ran Nielsen, besides putting into motion the transnational conspiracy from the UK and Hong Kong. A Howerstoke Trading director, identified as Niels George Vernon alias Robert Clifford, was suspected to have collected the Vessel Satcom terminal which was used by Nielsen and his associates for organizing the Purulia arms drop. In fact, while in Phuket after the arms drop, Nielsen called Vernon, who by that time had reached

London. He also called up one James William Bradford and D.H. Saunders of Sloane Avenue Mansions in London, besides Sovereign Trust International, the Hong Kong-based company that had helped Nielsen set up Carol Air Services Ltd in the Turks and Caicos Islands, and a certain Wan Tak San Dixon, most likely an employee of Wai Hong Mak.

While the Vessel Satcom terminal was leased in the name of Trevor Adams, the R&AW ascertained that two individuals with the same name visited India around the time of the Purulia arms drop. A Trevor Donald Adams, holder of a British passport (no. 000513221), issued in Liverpool on 10 July 1989, arrived at Dabolim Airport in Goa on 19 November 1995 by flight CKT 314 on a tourist visa, issued by the Indian High Commission in London on 30 October 1995. He stayed at Leela Beach Hotel and left on 3 December 1995 from Dabolim. Another person, with a similar name, Trevor George Adams, who held a British passport (no. 002703244) issued in Hong Kong on 20 December 1991, arrived in Delhi on 25 January 1996—a month after the arms drop—on an Emirates flight (EK 700). Like Trevor Donald Adams, his tourist visa was issued at the Indian mission in London on 5 December 1995. He put up at Delhi's Park Hotel before leaving India on 2 January 1996 by Indian Airlines flight IC 252 from Varanasi for Nepal. He returned to Delhi by flight IC 814 on 5 January 1996, before leaving for the UK two days later by Emirates flight EK 703.

What was found odd was that apart from the similar names, the date of birth of Trevor George Adams was 26 July 1935, and that of Trevor Donald Adams was 26 July 1945, i.e. exactly ten years apart. This gave rise to the suspicion that they were not only one and the same person, but aliases used by Niels George Vernon, alias Robert Clifford. It was a reflection of the tradecraft designed to protect the identity of anyone who may have assisted Nielsen in executing the Purulia arms drop plan. Two other names, Alexander Carrington and James Wilkinson, also believed to be fictitious, cropped up during the course of the investigation. It was later ascertained that both of these were aliases of Julian van Towsey, a former Australian army soldier who was associated with the Ananda Marga since the mid-seventies and was a subscriber to a telephone connection located to a

Copenhagen Street, London, address. However, Towsey was not found to have travelled to India on or about the time of the arms drop. As mysteriously as it had originated in February 1991, Howerstoke Trading Ltd wound up in May 1998, leaving little trail of some of the clandestine operations, including gunrunning, money laundering, smuggling of gold and precious stones, and drug trafficking, that it was deeply involved in.

In the course of a nearly four-hour-long interview in November 2012, an ex-R&AW head identified three features of the arms drop case—manifestations, as they are often described in intelligence jargon—that appeared 'interesting' to him. The first was Nielsen's links with the CIA-backed SPLA[53]. The second 'manifestation' was the unabashed protection that Danish authorities, including the police and the PET, provided to Nielsen for eight years before the Danish Ministry of Justice decided to extradite him to India. He was allowed to wander freely in Denmark despite an Interpol Red Corner Notice which Copenhagen, as a member state, was under obligation to execute against an internationally wanted terrorist. Related to this was Denmark's unenviable history of being a handmaiden of the CIA, which routinely used Danish airspace for its rendition flights to transport suspected Al-Qaeda and other Islamist jihadists to third countries for extra-judicial interrogations. The third was the strange circumstances under which the AN-26 aircraft was serviced at Karachi's Qaid-e-Azam Airport by Shaheen International, which had suspected links with a CIA front aviation company called Evergreen International. I will deal with each of these features and loose ends that I came across in government files and documents, 'seeping doubts and suspicion'[54] of some sort of covert operation that could throw light on how far and deep the Purulia arms drop conspiracy might have extended. As one trusted source and friend in the Russian Sluzhba Vneshney Razvedki (Foreign Intelligence Service), told me in December 2012, 'It was a quixotic, foolish operation.'

A 'Dirty Asset'

As Indian intelligence struggled to unravel the web of deception, the CBI went back to a clue that Bleach had provided while in the investigating agency's custody, that Nielsen had links with the SPLA. Bleach's initial suspicion of a Sudan connection was aroused when Nielsen insisted while in Karachi that the arms drop must take place on 17 December 1995, even if it had to be done in the darkness of night. Nielsen supposedly told Bleach that no delay of the operation was acceptable since he had 'other commitments' in Sudan which he was obliged to carry out.[55]

That Nielsen had links in Sudan, particularly with the SPLA, was confirmed after the CBI found a Filipino geologist's report on gold and gemstone mining in his briefcase. In the course of his travels to some African countries, especially Sudan, Nielsen developed strong contacts with leaders of the SPLA. A 25 September 1995 letter originating from a Calgary-based gas and oil exploration company called Altagem Resources Inc. recalled Nielsen's association with the then SPLA Chairman and Commander-in-Chief Dr John Garang de-Mabior, and agreed to enter into a gold-mining collaboration with the SPLA in South Sudan, besides seeking some assistance in Nairobi, Kenya.[56] At the height of the civil war in Sudan (between the SPLA and government forces) when the CIA supplied regular covert shipments of arms and 'non-lethal' military equipment to the primarily Christian rebel outfit, Nielsen was strongly suspected to have contributed to the Sudanese rebel forces' arsenal by supplying weapons in exchange for access to the gold mines in areas controlled by the rebel outfit.

A voluminous geologists' survey report, prepared by a Filipino mining expert and consultant, Jaime Zafra, found in Nielsen's briefcase, confirmed investigators' suspicions of his deep involvement in clandestine operations other than gold smuggling in Sudan. Zafra's LinkedIn profile claims he 'incorporated J.C. Zafra Geological Exploration & Operation Services, Inc. (1989 to present) with known colleagues and with clients Howerstock (sic) Ltd of Hong Kong for its Sudan, Kenya and Uganda gold and gemstone projects'[57].

Seventeen years after he met Nielsen, Zafra clearly recalled the 'two principles' who had approached him on behalf of Howerstoke Trading Ltd in the middle of 1995 to conduct a geological survey of the South Sudan, where they 'were looking for opportunities to extract gold, especially in the Kapoeta region'[58]. Zafra, who was not aware that Nielsen is now a globally wanted terrorist responsible for the arms drop over Purulia, had more to say, 'One of the principles from Hong Kong was of pure Chinese origin [Wai Hong Mak]. The other man, who introduced himself as Peter, said he was from the Netherlands.' Zafra continued, 'Since the SPLA couldn't pay in cash, the rebels would insist on paying in gold and rubies for the weapons that they received from Johnson [Nielsen].' When I asked which parts of Sudan did he conduct the geological survey, Zafra said, 'Almost the whole of South Sudan, especially Kapoeta.' In an email on 25 October 2012, Zafra said,

> The two met me in Hong Kong before and after fieldwork in South Sudan. They just sent with us a Singaporean to take charge of logistical support up to the site. The two did not mention to me about supplying weapons. It was only the men at the [survey] site who indirectly narrated me the story, but nothing was confirmed and I was then not interested in knowing because I was just doing a strictly professional geological consulting job. When I learned (though unofficially) of the arms trade in kind with minerals, I did not suspect their motives or objectives, except for the trade business that they have to pay in kind which for me made sense and [was] appropriate at that time.[59]

The potential recovery of gold in South Sudanese mines was pegged at an estimated 12 million ounces. The Kenyan government's Department of Mines and Geology provided an assay certificate (which vouches for the quality of extracted ore or metal) to Howerstoke Trading Ltd for the gold samples extracted in South Sudan's soil and stream sediments.

In a subsequent email, Zafra said that the Singaporean who went along with him to South Sudan, where he worked at the site for five days rather than two weeks as was agreed initially, was of Indian origin and sported a beard, which is a near-accurate picture of Deepak Manikan (a resident of Bedok Road in Singapore). In one of the meetings with 'Peter Johnson', as he knew Nielsen, Zafra recommended mining explorations in Kenya and Uganda, countries bordering South Sudan, for their potentially rich mineral deposits. Incidentally, some of the weapons that the CIA clandestinely supplied to the SPLA would be routed through Kenya and Uganda, whose governments wholeheartedly supported the SPLA cause. Indian authorities, especially the R&AW and the CBI, were left wondering whether Nielsen was a CIA 'dirty asset', supplying weapons to the SPLA on a freelance basis and in covert operations that would have all the elements of deniability. After all, the 'covert world is one of "smoke and mirrors", where it is not always easy to differentiate between men who are under contract to the CIA, work for "cut-out" corporations or are merely "freelancers for democracy" – mercenaries of fortune with a taste for adventure.'[60] In that sense, Nielsen was a true bill buccaneer.

While Indian intelligence and the CBI were wary of admitting Nielsen's role as a 'dirty asset', an outspoken British parliamentarian in the late nineties and early 2000s, Conservative MP representing Rochford and Southend, East, Sir Edward (Teddy) Macmillan Taylor, was quite unambiguous about Nielsen being a CIA agent. In a House of Commons debate on 27 November 2002, Sir Teddy said with characteristic wit and sarcasm:

> ...he has been travelling around quite a bit – he has been particularly active in Sudan – and that he was quite recently in London at Smith's hotel, where he was allegedly in the company of some rather official and well dressed people who are often seen at that excellent establishment. It was there that he allegedly sought to persuade one of my constituents, who resides in Great Wakering and whose details I would be glad to give to the Minister, to engage in some interesting work in Bosnia.
>
> Obviously, I do not expect the Minister to comment on these unusual events, but I would like him to say whether the Government regard[s] Mr. Kim Davey [sic] as a person who should be seized and questioned by the authorities, and whether approaches have been made to the Danish Government. The issue is of some significance. It seems that

although everyone says that they want to seize this gentleman, nothing seems to happen and he travels around doing rather interesting things.[61]

On a more serious note, Sir Teddy said that Nielsen would often be seen in the company of American and British intelligence officers in Nairobi (most notably in two hotels, Sarina and Safari Club) from where the operatives would monitor and support the covert war against Khartoum.[62] Not just Sir Teddy, other British sources, including Christopher Hudson, for long a Conservative politician and now a councillor in Suffolk county, recalled his 'friends in the (British) security establishment' sharing with him the information that Nielsen 'was a CIA rogue agent' whose 'connections my friend Peter Bleach had wholly underestimated. We know for a fact that he was trained by the Americans'. Hudson, who visited Calcutta twice after the arms drop to meet Bleach at Presidency Jail, was unsparing of British intelligence, saying that the security agencies 'had full knowledge' of the men who were behind the operation and who the weapons were intended for. 'An MI6 officer had followed Nielsen all the way to Calcutta in the third week of September 1995 when the Dane, a bad machine, a true Manchurian Candidate, was a guest at a local hotel. It was also not a secret to British intelligence that Nielsen was deeply involved in gold mining and oil prospecting in South Sudan,' [63] Hudson, whose memory and specific details of the arms drop did not appear to have dimmed in all these years, told me in a lengthy telephone conversation.

Protection at Whose Instance?

The Danish judiciary's stand on Nielsen was a body blow for the CBI. But that should not have come as a surprise for the investigating agency. Ten years before the Danish Ministry of Justice decided to start extradition proceedings against Nielsen, an Interpol officer had expressed dark forebodings of what the CBI could expect of Copenhagen. In an August 2000 email to a colleague, a senior Interpol officer deeply involved in the investigation to unravel the Purulia arms drop mystery, wrote, 'No officer [in Britain] talks to me concerning the Purulia case...I have a strong feeling that the Brits and the Danes do not want to see N [Nielsen] arrested.' One American special agent of the Bureau of Alcohol, Tobacco, Firearms and Explosives (ATF), Donald Manross, who was indoctrinated into Interpol's Purulia arms drop investigation, told me during the course of my inquiries sometime in November 2012 that 'neither MI5 nor MI6 officers would talk to us. They were reluctant to part with information.'[64]

Eight years prior to the August 2000 email, the same Interpol officer had cautioned his colleague in another email, saying, 'Do not forget how the Danes treated the Haestrup/Thune case, setting up all their demands in order to "protect" them.'[65] By the middle of May 2002, it was rather clear to the same Interpol officer who wrote another email to a colleague in India, 'I am sure the Danes will not extradite Nielsen so easily.' That prediction was to prove prophetic, especially in the backdrop of Denmark's claim—just before the CBI sought Nielsen's extradition to India—that it had not received the 1996 Letters Rogatory which detailed the central role he played in the arms drop. In 2002, the CBI, desperate to get Nielsen in its custody in the face of open protection accorded to the gunrunner by his country's police and security agencies, informally prodded Interpol to seek American help so that Washington could apply pressure on Copenhagen to extradite Nielsen to the US for the crimes he had committed there. This move came to nought. The Americans did not intervene. And why should the US act to apply pressure on Denmark, especially when the CIA and the PET enjoyed a very close working relationship that was revealed during investigations into the Iran–Contra affair in which Danish cargo ships were used to clandestinely transport huge quantities of weapons from Israel to Iran?[66]

Years later, events surrounding Denmark's acquiescence to US rendition flights that used Danish airspace to transport suspected Al-Qaeda and other Islamist jihadists to third-world countries for interrogation—that used torture as a means to extract information—would provide the cue to links between the CIA and the Danish PET. An Amnesty International report of 2007 pointed out that 'the Danish Government has, in a letter to the European Parliament's

Temporary Committee on the alleged use of European countries by the CIA for the transportation and illegal detention of prisoners, reported over 100 flights through Danish airspace and 45 stopovers in Danish airports by planes allegedly used by the CIA, including for renditions.'[67] The report observed that while in late 2005 the Danish government 'announced that unauthorized CIA flights would not be allowed into its airspace...the government rejected the recommendation to investigate and refutes the need to look into related legislation.'[68] There is, of course, no smoking gun evidence to indicate that the Danish authorities acted to protect Nielsen on the request of any Western country, but Indian diplomats who served in Copenhagen in the mid-2000s have privately narrated to me the possibility that PET was under some kind of influence to prevent Nielsen's extradition to India. This assumption is strengthened by reports that while the Danish government as well as the judiciary showed scant regard for the human rights of terrorism-related undertrials while they were in 'prolonged solitary confinement during the pre-trial phase of criminal proceedings'[69], they refused to extradite Nielsen on the grounds that he would face cruel, inhuman and degrading treatment, including torture, in an Indian jail. Clearly, there was something special about Nielsen, whose protectors feared he might reveal the nature and extent of his past relationship with them. 'There is strong possibility that a foreign intelligence agency used him and therefore the need arose to protect him,' retired R&AW Additional Secretary Bahukutumbi Raman said.[70]

Shaheen International's Background

Indian investigators and other senior officials believed, as early as 1996–97, that even though no Pakistani official or agency was involved in the arms deal or drop, there was reluctance on the part of Islamabad to share information, especially related to Shaheen International, which serviced the AN-26 and extended hospitality to the crew members. Regardless of Peter Bleach's claims that Shaheen had no links with the Pakistani security establishment, it emerged that Shaheen had links with a CIA front company called Evergreen International. Besides, Shaheen International was a subsidiary of Shaheen

Foundation run by some senior retired Pakistan Air Force (PAF) officers. Shaheen International was headed at that time by Javed Hayat, a retired PAF AVM. The other subsidiaries of Shaheen Foundation were Shaheen System, Shaheen Airport Services and Shaheen Airlines, which too were headed at the time by retired PAF officers. Besides, the PAF held 51 per cent of the shares of Shaheen Airlines. The rest of the equity (40 per cent) was controlled by M.I. Akbar, chief of the Akbar Group of Industries, and 9 per cent by Air Commodore (retired) Jamshed Akbar, a former Pakistani defence attaché in Moscow.

In February 1995, Shaheen Airlines introduced an international passenger flight between Peshawar and Dubai after hiring an Airbus 300 on 'wet lease'[71] from the US Continental Airlines. Before being inducted into the completely uneconomical route, the aircraft was serviced in the hangars of Evergreen International which, my sources in Indian intelligence said, carried out several clandestine operations of the CIA in the late seventies and early eighties. Evergreen International purchased a number of assets from the CIA, including helicopters, and its centre was at the base of Intermountain, another CIA proprietary, at Marana in Arizona.[72]

When the aircraft remained parked at Karachi airport on 22 November 1995, Nielsen showed great interest in developing business relations with Shaheen International. According to a concept paper on the agreement between Shaheen and Nielsen, the latter's company, Carol Air Services Ltd, was to be responsible for supplying aircraft at the best rates. The objective was to initially base one AN-26 at Karachi airport and subsequently 'bring in' bigger aircraft on a lease basis, with the 'intention' of bringing in at least an Antonov-12 to be based at Karachi as well. The flight routes contemplated were Karachi–Dhaka–Karachi, Karachi–Sharjah–Karachi, Karachi–Tashkent (and other cities in the Commonwealth of Independent States)–Karachi, and Karachi–Singapore–Karachi. On the other hand, for as long as the AN-26 remained stationed at Karachi airport, Shaheen International kept a close watch on the plane and its crew. There were at least two witnesses, the Russian ground engineers, Alexander Lukin and Vladimir Ivanov, who claimed that 'the staff of the airline,

Shaheen International, followed us to the aircraft when we conducted technical maintenance'[73].

Fifteen years after he was drafted into Interpol's Project Purulia, along with his colleague in the ATF, Victoria Lester, details of the investigations carried out in the US and Europe eluded Donald Manross when I first spoke to him over telephone on 17 November 2012. 'I lost contact with what happened to the key players after my departure from Interpol and retirement from service in 2000. There were so many different players. It was some sort of international conspiracy and we could track down the money trail in Hong Kong,' Manross said as he tried hard to rack his memory. 'There were so many intertwined fingers. It could have been one of two things. It was either the work of a gang of international smugglers or some government intelligence agency.'[74]

Although some R&AW officers like Jayant Umranikar did not 'find any direct involvement of the CIA' because of lack of any direct or visible links, it was very likely that 'international smugglers like Niels Christian Nielsen and some of his associates in the Ananda Marga were used as cut-outs to execute the covert arms drop operation.'[75] Shying away from actually identifying who the arms were intended for, Umranikar rued about 'some Tibetan activity' that the R&AW had noticed (in the form of some vague telephone and radio intercepts of conversations related to reviving links with the CIA) at the time, linking a suspected mid-nineties stirring with the armed rebellion of the Khampa tribesmen that the CIA had tried to fan in the fifties and sixties of the last century with Indian assistance. Umranikar asserted that the Purulia arms drop was a 'second dry run (following Nielsen's November 1995 Varanasi visit) and not the main operation which would have happened at a later stage, depending on the success or otherwise of the Purulia operation'[76]. I, however, believe the weapons that constituted the Purulia consignment was too large to be used for a 'second dry run'.

One of Umranikar's senior colleagues, who retired as a special secretary and had handled diverse country desks in the R&AW, was more circumspect about the Tibetans being the intended end-users. According to him,

The Ananda Marga as an organization was most certainly not the end-user. True, some of the Ananda Marga monks were used and it is also likely that the weapons cache would have been in safekeeping for sometime at the two-storeyed white building (in Bansgarh near Ananda Nagar) before being despatched elsewhere, though not among the cult's branches across the country. In which case, an entirely different cut-out would have been entrusted with the operation of ferreting out the weapons from the Bansgarh hideout, a move that, though fraught with risks, was not impossible.

'It could be that the objective was to keep things on the boil on the domestic front, create instability across the country and fuel disturbances as was the experience of several Latin American and African countries in the eighties and nineties,' he said, hinting at the involvement of a Western intelligence agency.

The R&AW special secretary's colleague in the IB, Ajit Doval, who rose to head the internal security agency in 2005 and in 2014 was appointed national security advisor to Prime Minister Narendra Modi, expressed scepticism about the theory that the Ananda Marga was the end-user. Doval, with years of counter-espionage and counter-terrorism special operations experience in Punjab, Jammu and Kashmir, and India's Northeast behind him, said,

> A group that intends to be the recipient of such a large consignment of weapons that were air-dropped over Purulia would have to have the capacity to absorb such a huge quantity of arms, including anti-tank grenades and rocket launchers, political and military will to take on government security forces, the logistics, secret training facilities and above all committed men to be part of an armed insurrection that could be sustained over a period of time. At the time of the arms drop, the Ananda Marga did not quite qualify on any of these characteristics.[77]

The weapons' quantity vastly exceeded the Ananda Marga's ability to organize, man and train the most hardcore of its members active in

Ananda Nagar at the time. The weapons-to-manpower ratio was far too much. Nielsen's public acknowledgement that the weapons would have been used to arm three groups of seventy-five men each as a vigilante force to 'protect' Ananda Marga establishments in Purulia from the onslaught of communists in the district was, according to Indian intelligence officers, a 'good' cover story to conceal the identity of the real end-users. The arms drop operation was by no means as simple as it was made to appear. Doval reasoned,

> If Nielsen believed that he could create a band of armed vigilantes who would freely move around Ananda Nagar with illegally procured AK-47 assault rifles and rocket launchers slung on their shoulders and yet go unchallenged by the police administration, he was then not quite aware of the dire consequences of taking on a duly elected state government. To do so would surely have invited disproportionate use of armed force by both the West Bengal and central governments, a consequence which he was surely fully aware of. Retribution and consequent legal action would have been swift.[78]

Nielsen's story before the Danish courts, therefore, was primarily aimed to project his 'cause' as political and not an act of international terrorism.

The Kachins

As I grappled with the different possibilities about the end-users, I pored over the files, looking for fragments that I might have missed. A thorough search of the enormous mass of documents yielded a curious piece of evidence in the incomplete jigsaw: for some reason, Yangon (Rangoon)[79], the former capital of the military junta-ruled Myanmar, figured unusually and very prominently in Nielsen's scheme of things, a path that the CBI appeared to have missed entirely, or chosen not to tread, in the course of its investigation. At the heart of this discovery were two provisional flight plans of the AN-26 aircraft. In the first, the aircraft was to take off from Karachi for Sharjah on 8 December 1995. After an overnight stay in Sharjah, it would fly to

Burgas, Bulgaria, the following day, with a technical stop at Turaif in Saudi Arabia. On 11 December the AN-26 would leave Burgas for Karachi with technical stops at Siirt in Turkey and Yazd in Iran. The next day, it would fly from Karachi to Yangon with technical stops at Varanasi and Dhaka. According to the flight plan, on the advice of Peter Bleach to the UK-based Baseops Europe, the aircraft would take off from Yangon for Karachi on 14 December with technical stops at Dhaka and Varanasi (see Chapter 5). Bleach's suggestion to Baseops was that the 'purpose of the visit to Rangoon is delivering passengers for business meetings'[80]. The second flight plan, this time prepared by Baseops Europe as late as 7 December 1995—ten days before the actual arms drop—retained Yangon, where the AN-26 would remain at the airport for two days, the longest stopover at any place in the course of its entire flight from Burgas to the Myanmar capital. Phuket, where the aircraft finally flew to after the arms drop, was nowhere in the plan. The second flight plan detailed:

ETD	KHI	12DEC0500Z
ETA/D	VNS	12DEC0730/0830
ETA/D	DAC	12DEC0945/1045Z
ETA	RGN	12DEC1230Z
ETA	RGN	14DEC0230Z (0900LT)
ETA/D	DAC	14DEC0420/0520Z
ETA/D	VNS	14DEC0635/0735Z
ETA	KHI	14DEC1000Z[81]

The penultimate flight plan underwent substantial changes, owing primarily to a three-day stopover at Karachi airport where the aircraft, loaded with the weapons consignment, arrived on 13 December and finally departed for Varanasi on 17 December. Before taking off for Varanasi, the aircraft had not received permission from Dhaka to land at Zia International Airport and the Myanmar authorities had 'acknowledged, but not granted'[82] permission to land at Yangon. One reason why Yangon refused to grant landing permission to the AN-26 was that Baesops Europe, which coordinated all the landings and take-offs and other flights details for Nielsen and Bleach, had not given the authorities the name of the sponsors or the purpose of the flight.[83] Nielsen had not shared that information, whatever its nature, with Baseops. It was at that time that Nielsen thought up the idea of flying to Phuket, although he knew full well that the distance from Varanasi to the Thai sea resort town could not be covered without a refuelling stop at the Myanmar capital, where the authorities ultimately did not give landing permission even after the flight took off from Varanasi. The Civil Aviation Authority of Bangladesh, too, had denied permission to land. It was a combination of luck, fortitude and the derring-do of Nielsen—besides the ignorance of the authorities in Calcutta, who had no clue that the AN-26 had disgorged nearly five tons of lethal weapons before trying to leave Indian airspace—which allowed for refuelling of the aircraft at Calcutta's Dum Dum airport in the early hours of 18 December 1995.

Why was Yangon so important for Nielsen? A technical stop at Yangon for refuelling would not have taken more than an hour, yet the original plan was for a two-day stopover in Myanmar's capital. If the Myanmar authorities had actually given permission for landing, what would Nielsen have done for two days in Yangon, especially when Phuket was not even part of the original flight plan? Yes, the Latvian crew would have required the much-needed rest before labouring again to fly back the same route to Karachi via Dhaka and Varanasi. But Yangon would have been an unlikely place to celebrate the arms drop or give the crew members the vacation they did ultimately enjoy in Phuket. In his book, *De Kalder Mig Terrorist* (They Call Me a Terrorist), Nielsen avoided naming Yangon as his original destination. Was Nielsen scheduled to meet someone in Yangon if the plan had proceeded unhindered? Was it to be a representative of an insurgent group fighting the military junta in Myanmar? The alleged arms drop financier, Wai Hong Mak, rushed to Phuket to meet Nielsen with $20,000 in cash on 19 December 1995.[84] Would Mak have done the same had the aircraft been routinely allowed to land at Yangon? After parachuting the weapons over Purulia and overflying Calcutta, Nielsen agreed with the AN-26's pilot Alexander Klichine to brazen it out and try to land in Yangon, citing emergency refuelling.[85] The plane flew over the Bay of Bengal on its way to Yangon, but over Dopid (89.12 degrees east and 12.55 degrees north), a pre-fed waypoint in the GPS, the pilot realized there was not enough fuel to take the aircraft to Yangon and returned to Calcutta to refuel. Subsequently, the landing at Phuket, in the backdrop of the botched Purulia arms drop operation, was an emergency arrangement. By that time, Nielsen had foreseen trouble ahead and had ordered for replenishment of contingency funds, which were supplied by Mak. Of course, Phuket also served to give the Latvian crew members a happy and warm vacation by the Gulf of Siam, especially after they learned, much to their chagrin, that the flight from Burgas was intended to airdrop arms to an insurgent group.

A second, curious piece of evidence, which indicated that the arms were perhaps destined for a Myanmar insurgent group, appeared in the form of two sets of coordinates marked on two maps 'in foreign language'[86] embedded in the tome of incriminating documents that the CBI seized from the AN-26 after it was impounded at Mumbai airport on 22 December 1995, but never paid any attention to in the course of its investigation. The first set of coordinates was 42.60 degrees east and 4.16 degrees north, which is close to the tri-junction of northeastern Kenya, Ethiopia and Somalia. The second set was 96.11 degrees east and 28.40 degrees north. I sat puzzling over these coordinates. On the face of it, they made no sense, just numbers carelessly typed out by a CBI clerk as part of the agency's own inventory of documents. But when I used Google Earth to check and verify the coordinates, with assistance from a retired IAF officer, the finding was unmistakable and startling. The coordinates indicated a point around the tri-junction of Tibetan China, Arunachal Pradesh and Myanmar's Kachin State. In fact, the coordinates pointed to an area just northeast of the Kachin State border that juts into China like a blunt horn. A magnified image of the point on Google Earth showed the place to be completely devoid of human habitation, mountainous and heavily forested—an ideal location for an arms drop in an inhospitable and rugged terrain, where it was not impossible for the fierce and determined Kachin insurgents to penetrate and grab weapons landing from the skies above. Apprehension had become a tangible piece of evidence, but that is where the paper trail ended.

Questions swirled in my mind. Were the original coordinates—

86.20 degrees east and 24.30 degrees north—which were discussed and charted on a map by Nielsen, Peter Bleach, Wai Hong Mak, Randy and Peter Jorgen Haestrup at the 27 September 1995 meeting a red herring? Was it one of several pieces of disinformation[87] that Nielsen provided for Bleach's consumption? Or was Purulia Plan B because the Bangladesh and Myanmar authorities had refused landing permission to the AN-26? Did the plan include landing a weapons-laden aircraft at Dhaka's Zia International Airport, from where it would fly north, para-drop the arms cache close to the Kachin State border with China and then return south to land at Yangon?

If the arms were indeed meant for the Kachins who, at the time, controlled a vast swathe of territory rich in jade, gold and rubies[88], but had been dropped over Purulia, how were they to be transported from the West Bengal district to as far away as Myanmar? The risks of detection would be enormous and might even lead to an operational debacle, assuming that the arms drop over Purulia went undetected in the first place. Yet, 'That sounds plausible,' was the response to my questions from a former director of the IB, who otherwise had no doubts that the arms drop was a covert operation by a foreign intelligence agency of a Western country with foreign policy objectives in Myanmar, that had begun to drift close to China in the early nineties. According to this former IB chief, an ardent advocate that the IB, rather than the R&AW, ought to have long ago been given charge of intelligence gathering and covert operations in India's near-abroad (neighbouring) countries, several clandestine means could have been employed to transport the weapons to the Kachins via India's Northeast—shipped by overland route or concealed in trucks or smaller vehicles in small quantities till as far as the international border that Nagaland and Manipur share with Myanmar.

The terrorist and insurgent groups active in South Asia in the early and mid-nineties—the mujahideen guerrillas in Afghanistan; the Taliban, Arab, Afghan and Pakistani terrorists who operated in Kashmir; the LTTE in Sri Lanka; the Naga, Assamese (ULFA) and Manipuri rebels in India's Northeast; and the KIA in Myanmar—displayed, in varying degrees, tremendous fighting capabilities coupled with the will to sustain themselves militarily and politically. In the words of the former IB director, Ajit Doval, they had the 'capacity to absorb' weapons of the kind that were para-dropped over Purulia and use them effectively, sometimes with ruthless efficiency. In 1995, however, the KIA rebel army was into the first year of a ceasefire agreement with the Myanmar military. The cessation of hostilities in February 1994[89] was preceded by a series of military debacles and miscalculations on the part of the KIA as well as its political wing, the Kachin Independence Organisation (KIO). One of the main tactical miscalculations of the KIA was to send a delegation to Pakistan to meet with the 'cold and ruthless'[90] Afghan mujahideen leader Gulbuddin Hekmatyar, with the objective of procuring American Stinger missiles. But the 'Indians found out about it and stopped the deal'[91]. A second mistake the KIA made was to disregard the R&AW's express suggestion that it must not attack Myanmar army camps near the Indian border 'which would jeopardise supply lines and alert the government in Rangoon'[92]. But, the KIA, flush with weapons supplies from the R&AW in the late eighties and early nineties, went ahead anyway. Although the KIA rebels were able to overrun and capture two Myanmar army camps at Pangsau, close to the India–Myanmar border, and Namyung, the two critical outposts were retaken[93] by the end of 1992 when the Indians stopped providing weapons to the Kachins.

The R&AW was not the only source of arms for the KIA. In 1986–87, the KIA attempted to procure a major consignment of arms from China. Some of the rebel leaders contacted a group of ethnic Chinese in Hong Kong (at least one of them was an Indonesian-Chinese) who claimed they would be able to purchase a large quantity of weapons from the Chinese parastatal, NORINCO[94]. The deal was to be made through what was said to be a North Korean front company based in Tokyo. The group also claimed to be working for T.K. Ahn, a fairly well-known Hong Kong tycoon. The KIA paid anywhere between $10 and $20 million for the shipment, but were then told that it would not be possible to deliver the weapons across the border into Myanmar's Kachin State from China. The Chinese supposedly did not agree to that. So the Kachins turned to other possibilities, including approaching the R&AW station chief in

Bangkok, pleading for and requesting the agency's assistance for the safe passage of the weapons' consignment through Indian territory in the Northeast. The KIA request, made sometime in late 1989, was flatly turned down by the R&AW station chief. Around the same time, a source close to the Kachin rebel leadership said that the rebels 'were discussing the possibility of an airdrop in Nepal'. The guns were to be flown to an airfield in Nepal from where smaller aircraft would fly over Kachin State and drop the weapons there. But the plan never took off.

In the end, the Kachins did not get any weapons from the Chinese gang, which pocketed the money they had received from the KIA leadership. The Chinese authorities held an inexplicable press conference in Beijing, saying they would never allow their territory to be used for the transfer of weapons to 'friendly neighbouring countries (read Myanmar)'. At least one of the Indonesian-Chinese involved in the 'deal' with the KIA was arrested in China and then disappeared without a trace. The whole story, according to my source, was hushed up. As a consequence of the bad deal, the then KIO chairman, Maran Brang Seng, offered to resign but was persuaded to stay on. When I tried to verify the account with R&AW officers who handled and supervised operations in Myanmar from Bangkok, I met with stony silence.

Due to battlefield reverses in the early nineties, followed by China's growing warmth with Myanmar and India's sudden coldness toward the Kachins, the KIA leadership turned to Western sources for the supply of weapons. A senior, septuagenarian 'officer' in the KIO feigned loss of memory when I asked him about the different sources of arms, though he did admit that most insurgent groups in Myanmar were supplied military equipment by Western intelligence agencies in the early and mid-nineties. Christopher Hudson, however, recalled that 'it was Peter [Bleach] who had shared with me his initial suspicion that the Purulia consignment was meant for a Burmese insurgent outfit'. Elsewhere, Bleach had expressed strong suspicion that the weapons were intended to be dropped 'somewhere further east'[95]. Hudson recalled a most intriguing incident when he was in Calcutta in early 1996 to meet Bleach in Presidency Jail:

One afternoon, a young man, who appeared to be of mixed race, dropped by unannounced at Taj Bengal Hotel, where I was putting up, and delivered a calendar in the name of a Burmese insurgent group (either the Karen National Liberation Army or the Kachin Independence Army). The front cover picture, set in the backdrop of mountainous jungles, was of armed young men in their customary battle fatigues and berets. At the time, it was a strange memento. It did not strike me then as something important, but when I later reflected about the calendar's significance, it seemed to make sense. The calendar was a signal which indicated that Nielsen had earlier worked with at least one of the Burmese insurgent groups. According to Peter Bleach, Nielsen was earlier involved in smuggling jade out of Burma besides being part of an illegal trade in cigarettes in China.[96]

A source in a West European country police force familiar with the arms drop case told me that 'Burma was most definitely discussed' among CBI and Interpol investigators. 'From the beginning, Burma was on the table and the CBI as well as other [Indian] security officials discussed the [political] situation in that country and the relative strengths and need for weapons of the some of the insurgent groups that operated there at the time. However, there was no documentary or oral evidence to pinpoint a particular rebel outfit in Burma that could have been the end-user,' the officer, who spoke on the express condition that his name would not be revealed, said in the course of a telephone interview. If the investigators had plumbed Peter Bleach, some information related to Nielsen's smuggling activities in Myanmar could have emerged.

Conclusion

This is as far I could go with this immensely vexing, extraordinarily complex and tangled, but absorbing and enjoyable case. There were really only two courses open to me to investigate a case such as the Purulia arms drop: to make intelligence guesses based on the evidence and fragments of intelligence and to analyse where they led to. The second was to adopt a more holistic approach, to analyse every available piece of information carefully and with some degree of precision, so that a plausible explanation could emerge. With limited resources at hand, I tried to integrate both, though, admittedly the second approach lacked scientific rigour, leaving some, if not all, of the questions unanswered.

I am convinced that the R&AW and the IB have a central role to play if the security and the political establishments ever decide to pursue the loose ends which would help them get to the bottom of the conspiracy. For all their faults, weaknesses and other drawbacks, the security agencies together possess a rich trough of information and have the capabilities to complete the Purulia jigsaw. If the security agencies do know—which I suspect they most certainly do—the ring of men who conspired to plan and launch the covert operation, if not the intended recipients of the weapons, they must be held accountable so that the air of suspicion and the doubts that have lingered for years is cleared, even if it means something of their hand being exposed. The two agencies must shed their bureaucratic baggage, adopt more than a professional approach, emerge from their passive shells out in the aggressive open and bring to light one of Indian national security's most enduring mysteries. To achieve that objective, they must, first of all, demonstrate willingness to revive the case, acknowledge the failures of the past and then expose the conspiracy and identify the real end-users. Otherwise, the intelligence agencies will have to forever live with the taint of failure, lethargy, dysfunction and incompetence.

Clearly, Nielsen knows a lot more than he has penned in his book (which is strewn with lies, half-truths and cover stories) or spoken over Indian television or the Danish media. I am certain, as were Interpol and CBI officers, that Nielsen enjoyed the protection of the Danish authorities, including the police and the PET. An Interpol officer went so far as to claim that 'he was connected with Danish intelligence'. Whether those 'friends' acted on behalf of other friends in a friendly foreign intelligence agency with which the PET has collaborated in the past, constitutes a crucial element in the entire Purulia arms drop case. In the final analysis, neither Britain nor India nor even Denmark and the US came out of this case with honour.

It is, therefore, incumbent upon the government to pursue every available means to bring Nielsen to justice in India. Any lack of interest or stonewalling, willing or otherwise, to force Denmark's hands to extradite Nielsen would only reinforce suspicions of an Indian government role in suppressing the truth from emerging. In the past, the BJP-led government of the National Democratic Alliance was found to be soft towards the Latvians and the British arms dealer, Peter Bleach. Instead of being apprehensive or terrified that further inquiries might trigger unsavoury revelations of the involvement of now-friendly foreign intelligence agencies, the government of the day must confront the conspiracy and expose the men and organizations behind the Purulia arms drop. While making loud noises over seeking the extradition of Nielsen may go down well with the public at large, the government must answer the question as to why no effort whatsoever, since at least 2004, has been made to hunt down three of the Ananda Marga monks; a Singaporean of Indian descent, Deepak Manikan; and the Hong Kong Chinese, Wai Hong Mak, who were involved in the conspiracy and the act of gunrunning. That little police or intelligence effort went into pursuing these fugitives is a telling comment on the agencies' apathy towards the case. I believe that the existing evidence and the loose ends that have been highlighted in the chapters of this book are potentially 'live' and must be followed up and pursued relentlessly, regardless of the unpleasant facts that might surface. A no-holds-barred investigation into every aspect of the case is a national imperative. The case must come to a closure— with the full truth.

NOTES

INTRODUCTION

1. Chandan Nandy, 'Indian mole may have fled to Colombo', *The Telegraph*, 31 December 1995, p. 1.
2. Chandan Nandy, 'Davy top gold smuggler: CBI', *The Telegraph*, 1 January 1996, p. 1.
3. Claiming to be a socio-cultural organization, the Ananda Marga was founded in 1955 in Jamalpur, Bihar, by P.R. Sarkar, alias Anandamurti. Sarkar projected himself as an incarnation of God and a halo of mystery developed around him. Many of his followers, in fact, believe that he is invested with supreme powers. Educated young men were selected by the organization as whole-time workers.
4. P.C. Sharma was CBI director from March 2001 to December 2003.
5. A brief history of the CBI is available at www.cbi.nic.in.
6. Cabinet Secretariat, 'Statement on behalf of the Government of India on the Purulia Arms Drop Case', Press Information Bureau, 30 April 2011.
7. In intelligence or espionage parlance, a 'cut-out' is a mutually trusted intermediary between two agents. A cut-out facilitates the exchange of information between two agents. He/she usually only knows the source and destination of the information to be transmitted, but is unaware of the identities of any other person involved in an intelligence operation. In other words, a cut-out is a go-between used to preserve the safety and anonymity of the principals.
8. For reasons of security and due to a request for anonymity, I cannot reveal the identity of the officer, who continues to serve a police force in the country.
9. Anirban Roy, 'Purulia arms drop man to 'grace' Vijay Diwas event', *The Bengal Post*, 15 December 2012, p. 1.

10. Nielsen possessed two forged New Zealand passports, one of which was 'issued' in the name of Kim Palgrave Davy and the other in the name of Kim Peter Davy.

I. COLLECTIVE IRRESPONSIBILITY

1. R&AW UO Note of 25 November 1995.

2. There is evidence that British intelligence provided early information to R&AW in three separate tranches: 10 November, 17 November and 15 December 1995.

3. Christer Brannerud, 'Project Purulia', Analytical Criminal Intelligence Unit, International Criminal Police Organization (Interpol), March 1999, p. 5. According to a decision of the Interpol headquarter, Lyon, the highly classified Project Purulia report was shared with the CIA, but not with the intelligence services of Britain or Denmark.

4. The Q Branch is responsible for liaising with friendly foreign intelligence agencies.

5. After the arms drop and the subsequent leak that intelligence had been shared by British intelligence, the MI5 SLO in the UK High Commission in New Delhi was in a major flap. An R&AW joint secretary who was 'in the know' of the developments at the time recalled that the SLO representative met then Indian Home Minister S.B. Chavan and Home Secretary K. Padmanabhaiah, among others, to not just express annoyance at the leakage but also to make attempts to steer clear of the controversy that was generated after it became public that the British Security Service had initially supplied R&AW with information about a possible arms drop in India.

6. E. Ahamed, Rajendra Agnihotri, Mukhtar Anis, Illyas Azmi, L. Balaraman, Dileep Singh Bhuria, Bhavana Chikhalia, Paban Singh Ghatowar, Ramkrishna Kusmaria, Sanat Mehta, Hannan Mollah, Jayanta Rongpi, C. Silvera, Tilak Raj Singh and Purnima Verma, 'Third Report Purulia Arms Dropping' Committee on Government Assurances, Lok Sabha Secretariat: New Delhi, 7 May 1997, p. vii

7. Ibid., p. 6

8. Ibid.

9. Ibid., p. 13

10. Ibid.

11. Ibid.

12. Ibid.

13. The Jain Committee, as it came to be known at the time, was chaired by Home Ministry Special Secretary (Internal Security and Police) V.K. Jain. In one of its early meetings after the arms drop, Jain's suggestion that the Purulia operation was the work of Russian intelligence raised several eyebrows, especially among senior R&AW officers.

14. R&AW note of 28 December 1995, prepared for a meeting of the core group the same day.

15. Ibid., p. 1

16. Ibid., p. 2

17. Ibid., p. 1

18. Report of the V.K. Jain Committee, 29 December 1995, p. 1

19. Ahamed et al., 'Third Report Purulia Arms Dropping', p. 14

20. Ibid.

21. Ibid., p. 23

22. Ibid.

23. Ibid.

24. Ibid., p. 24.

25. Contrary to belief that two IAF fighter jets were scrambled, an Indian Air Force Headquarter Note of 1 January 1996 revealed that IAF 'aircraft at Bhuj were armed and placed on ORP (operational readiness platform) for intercepting the aircraft, if the need arose.'

26. Ibid., p. 20.

27. L.K. Advani, George Fernandes, Jaswant Singh and Yashwant Sinha, 'Reforming the National Security System: Recommendations of the Group of Ministers', National Security Council Secretariat, February 2001, p. 58.

28. Thomas H. Kean, Lee H. Hamilton, Richard Ben-Veniste, Bob Kerry, Fred F. Fielding, John F. Lehman, Jamie S. Gorelick, Timothy J. Roemer, Slade Gorton and James R. Thompson, 'The 9/11 Commission Report: Final Report of the National Commission on Terrorist Attacks upon the United States', W W Norton & Company: New York, 2003, p. 400.

2. IS IT A CIA SCAM?

1. Brannerud, 'Project Purulia', p. 7.
2. Defining the terms 'irrevocable' 'confirmed' and 'transferable'—
   Irrevocable: The buyer cannot cancel the letter of credit once it is
   issued. It would, however, have a defined period of validity and
   would expire if not transacted during that period. Confirmed: The
   receiving bank (of the seller) contacts the issuing bank (of the
   buyer) and receives an independent confirmation that the funds are
   available and are set aside specifically to cover the letter of credit.
   Transferable: On receipt of the letter of credit, the seller can lodge
   it in his/her bank and use it to cover the letter of credit issued to
   the manufacturer. This is also known as 'backing off', which means
   that the banks are acting as 'honest brokers' between two business
   entities. The buyer cannot fail to pay the seller, provided that the
   seller actually ships the goods. Conversely, the seller cannot get the
   money unless he ships the goods. Both parties are, therefore, safe
   and the need for cash deposits, etc. is removed. This was standard
   practice in the defence trade as also in trading of other commodities
   in the West.
3. Peter J.G. Bleach, 'Delivery of Arms by Air to West Bengal',
   Statement to the CBI in Calcutta, January 1996, p. 13. Some parts of
   this narrative have been extracted from Bleach's statement to the
   CBI while in captivity. It is my belief that while writing out the
   fifty-page detailed statement, Bleach made a clean breast of things
   in the hope that a 'confession' at the initial stage of the CBI's
   investigation would stand him in good stead and that he would be let
   off by the Indian authorities for being cooperative.
4. Ibid., p. 14
5. Ibid., p. 15.
6. Born on 20 May 1951, in Halifax, UK, to an English father (Albert
   James Bleach) and a German mother (Oceana Felicia Joana
   Magdalena), Peter James Gifran von Kalkstein Bleach was educated
   at the Uplands Preparatory School (formerly St Peter's School) in
   Scalby, Yorkshire, between 1964 and 1969. He completed the 'A'
   level and after graduation enlisted in the British Army Intelligence
   Corps. He entered for training at HQ Intelligence Centre at Templar
   Barracks in Ashford, Kent. He passed corps training with distinction

and was awarded best recruit. In 1972, he was posted to 39 Airportable Brigade, HQ Northern Ireland, Thiepval Barracks, Lisbum, Northern Ireland, serving in the brigade intelligence section. Thiepval Barracks would later become the command centre for the British Army's Force Research Unit (FRU) whose members recruited and handled agents to infiltrate the Provisional Irish Republican Army (PIRA). Bleach subsequently passed all corps training courses and promotion examinations with distinction, rising to the rank of sergeant. He remained at HQ Northern Ireland until the end of 1972, leaving after a disciplinary offence led him to be offered transfer to another unit. The transfer was turned down. Between 1973–1981, Bleach travelled to Southern Rhodesia (later rechristened Zimbabwe) and joined the Rhodesia Prison Service. He took early retirement on pension after reaching the rank of assistant superintendent in 1981. Next, he spent three years as commanding officer, Chipinga station, on the Mozambique–Southern Rhodesia border, serving with Joint Operations Command for Operation Thresher. After several security-related jobs in Zimbabwe and Johannesburg, South Africa, Bleach returned to the UK in the middle of 1985 and until 1993 was a self-employed security advisor and a consultant in commerce and industry.

7. Bleach, 'Delivery of Arms by Air to West Bengal' p. 15.
8. Brannerud, 'Project Purulia', p. 7. This is among several others of Davy's Visa card transactions that the Interpol tracked to locate him in different parts of the world as part of its global manhunt for one of the most wanted suspects.
9. Ibid.
10. Cross-examination report of Brian Thune and Peter Haestrup prepared by the Copenhagen Criminal Investigation Department on 25 October 1995, p. 30.
11. Ibid., p. 32. Thune went underground immediately after the arms drop, but was arrested in August 1996 and remanded in custody of the Fredeikssund police on charges of fraud and forgery. His interrogation took place for several days in September 1996, in a Hillerod jail where he was placed in isolation.
12. Ibid., p. 10. When I cross-checked with Novo Nordisk's Hong Kong branch, I was told that there could not have been a CEO in the early nineties as the branch came into being only in 2002. In his statement

to the CBI, Peter Bleach described Flemming Soderquist as a 'Danish national resident in Hong Kong. Peter Haestrup claims that Sodequist is his colleague and business partner in Hong Kong. However, he may only be an acquaintance whose premises Haestrup makes use of. At Haestrup's suggestion, I have previously forwarded requests for commodity prices to Soderquist, but I did not receive any response.' (P. 7). Most likely, Flemming Soderquist was an undercover Danish intelligence officer who shadowed Bleach to find out more about his background.

13. Ibid., p. 36.
14. Ibid.
15. A broker in aircraft spares, Roeschke's company went by the name of Rofa Air Services. Bleach claims to have met Roeschke sometime around 1992–93 at an air show where they exchanged business cards. The two remained in fairly regular telephone contact and Bleach put a number of business inquiries Roeschke's way and vice-versa.
16. Bleach, 'Delivery of Arms by Air to West Bengal', p. 10. This was Bleach's normal practice when dealing with a new and unknown company.
17. A little over a month after Bleach sent the quotation to Haestrup, the final sixteen-page contract (No. 10026/95/A) was signed between the British arms dealer and Kim Peter Davy in London on 13 September 1995. Davy's gun of choice was the AK-47 Type 56 Mk-2.
18. Remember, this was 1995, barely six years after the break-up of the Soviet Union and the widespread chaos that prevailed in some of the Warsaw Pact countries, where weapons-manufacturing companies had plenty of unused stock.

3. LIKE VULTURES TO A CARCASS

1. Jayantha Dhanapala, 'Multilateral Cooperation on Small Arms and Light Weapons: From Crisis to Collective Response', *The Brown Journal of World Affairs*, Spring 2002, Vol. IX, Issue 1, p. 163
2. Denise Garcia, *Small Arms and Security: New Emerging International Norms*, Routledge: Oxon (UK), 2006, p. 4.
3. Report of the Panel of Government Experts on Small Arms, United

Nations Document A/52/298, United Nations, New York, 7 August 1997.

4. Jasjit Singh, 'Illict Trafficking in Small Arms: Some Issues and Aspects', in Pericles Gasparini Alves and Daiana Bellinda Cipollone (eds), *Curbing Illicit Trafficking in Small Arms and Sensitive Technologies: An Action-Oriented Agenda*, United Nations Publication: New York and Geneva, 1998, p. 13.

5. Brian Wood and Johan Peleman, *The Arms Fixers: Controlling the Brokers and Shipping Agents*, Joint Report of the Norwegian Initiative on Small Arms Transfers, International Peace Research Organisation and the British American Security Information Council, November 1999, p. iii.

6. Douglas Farah and Stephen Braun, 'The Merchant of Death', *Foreign Policy*, No. 157, November-December 2006, p. 47.

7. 'International Crime Threat Assessment', US Government Inter-Agency Working Group, December 2000, p. 32.

8. Wood and Peleman, *The Arms Fixers*, p. 1.

9. Andrew Feinstein, *The Shadow World: Inside the Global Arms Trade*, Penguin Books: London, New York, New Delhi, 2012, p. xxiii.

10. Stephanie G. Neuman, 'The Arms Trade, Military Assistance and Recent Wars: Change and Continuity', *Annals of the American Academy of Political Science*, Vol. 541, September 1995, p. 50.

11. Michael T. Klare, 'Awash in Armaments: Implications of Trade in Light Weapons', *International Harvard Review*, Vol. 17, Issue 1, Winter 1994–95, p. 24.

12. Feinstein, *The Shadow World*, p. xxiii.

13. For a detailed account of the linkages between narcotics and arms trafficking, see Tara Kartha, 'Southern Asia: Narcotics and Weapons Linkage', in Jasjit Singh (ed), *Light Weapons and International Security*, Pugwash, IDSA and BASIC: New Delhi, 1995, pp. 63–86.

14. Neuman, 'The Arms Trade, Military Assistance and Recent Wars', p. 65.

15. David Kinsella, 'The Black Market in Small Arms: Examining a Social Network', *Contemporary Security Policy*, Vol. 27, No. 1, April 2006, p. 104.

16. Michael T. Klare, 'Combating the Black-Market Trade', *Seton Hall Journal of Diplomacy and International Relations*, Summer/Fall 2001, p. 44.

17. T.R. Naylor, 'The Structure and Operation of the Modern Arms Black Market', in Jeffrey Boutwell, Michael T. Klare and Laura W. Reed (eds), *Lethal Commerce: The Global Trade in Small Arms and Light Weapons*, American Academy of Arts and Science, Committee on International Studies: Cambridge, MA, 1995, p. 48.

18. Dhanapala, 'Multilateral Cooperation on Small Arms and Light Weapons', p. 165.

19. 'Small Arms: Report of the Secretary General' UN Security Council, 17 April 2008, p. 4. For a comprehensive account of illegal trade and transfer of weapons, especially SALWs, see 'Dead on Time: Arms Transportation, Brokering and the Threat to Human Rights', Amnesty International and TransArms, 2006, p. 147.

20. Feinstein, *The Shadow World*, p. 435.

21. 'The illicit trade in small arms and light weapons in all its aspects', Note by the secretary general, United Nations General Assembly, 30 August 2007, p. 1.

22. Wood and Peleman, *The Arms Fixers*, p. 4.

23. Edward J. Laurance, 'Political Implication of Illegal Arms Exports from the United States', *Political Science Quarterly*, Vol. 107, No. 3, Autumn 1992, p. 502.

24. Peter Lock, 'Pervasive Illicit Small Arms Availability: A Global Threat', HEUNI Paper No. 14, The European Institute for Crime Prevention and Control, Helsinki, 1999, p. 9.

25. Wood and Peleman, The Arms Fixers, p. 12.

26. Frederick M. Kaiser, 'Impact and Implications of the Iran-Contra Affair on Congressional Oversight of Covert Action', *International Journal of Intelligence and Counter-Intelligence*, Vol. 7. No. 2, Summer 1994, p. 206.

27. Brian Champion, 'Subreptitious Aircraft in Transnational Covert Operations', *International Journal of Intelligence and Counter-Intelligence*, Vol. 11, No. 4, Winter 1998-99, p. 453.

28. Ibid., p. 470.

29. Laurance, 'Political Implications of Illegal Arms Exports from the United States', p. 508.

30. Stephen D. Goose and Frank Smyth, 'Arming Genocide in Rwanda', *Foreign Affairs*, Vol. 73, No. 5, September-October 1994, pp. 88–89.

31. Singh, 'Illict Trafficking in Small Arms', p. 13. Singh's emphasis is in italics.

32. Garcia, 'Small Arms and Security', p. 41.
33. Lock, 'Pervasive Illicit Small Arms Availability', p. 15.
34. Wood and Peleman, *The Arms Fixers*, p. 5.
35. Report of the Panel of Experts appointed pursuant to UN Security Council Resolution 1306 (2000), para. 19 in relation to Sierra Leone, December 2000.
36. Garcia, 'Small Arms and Security', p. 45.
37. UN General Assembly Report, 30 August 2007, p. 3.
38. Gurmeet Kanwal and Monika Chansoria, 'Small Arms Proliferation in South Asia: A Major Challenge for National Security', Centre for Land Warfare Studies, Issue Brief No. 19, May 2010, p. 1.
39. Wood and Peleman, *The Arms Fixers*, p. ii.
40. Ibid, p. 12.
41. UN General Assembly Report, 30 August 2007, p. 1.
42. 'India: Integrated National Report on the Implementation of International Instrument to Enable States to Identify and Trace, In a Timely and Reliable Manner Illicit Small Arms and Light Weapons in all it's aspects', Ministry of External Affairs, New Delhi, 31 March 2008, p. 5.
43. Ibid, p. 8.
44. Ravinder Pal Singh, 'Indian government at the United Nations Arms Trade Treaty negotiations', Control Arms Foundation of India, 3 July 2012. Accessed from http://www.cafi-online.org/articles.php?event=det&id=72 on 3 December 2012.
45. 'Taming the Arsenal: Small and Light Weapons in Bulgaria, South Eastern Europe Clearinghouse for the Control of Small Arms and Light Weapons', United National Development Programme, March 2005, p. 33.
46. PTI, 'India displaces China as world's largest arms importer', *The Times of India*, 19 March 2012. Accessed from http://articles.timesofindia.indiatimes.com/2012-03-19/india/31209934_1_sipri-arms-importer-international-peace-research-institute on 3 December 2012. See also Stockholm International Peace Research Institute (SIPRI) press release of 14 March 2011, at http://www.sipri.org/media/pressreleases/2011/armstransfers.
47. Prashant Diskshit, 'Why the Arms Trade Treaty matters', *Indian Express*, 23 July 2012. Accessed from http://www.indianexpress.com/

news/why-the-arms-trade-treaty-matters/977956 on 3 December 2012.

48. Prashant Diskhit, 'Addressing Indian Dilemmas on the Arms Trade Treaty', Institute for Defence Studies and Analysis, New Delhi, 15 May 2012. Accessed from http://www.idsa.in/idsacomments/AddressingIndianDilemmasontheArmsTradeTreaty_PrashantDikshit_150512 on 3 December 2012. See also Rahul Prakash, 'Arms Trade Treaty: An Indian Perspective', Observer Research Foundation, Issue Brief No. 39, July 2012, p. 6.

4. 'THE INDIANS COULD BE ADVISED OF THE SITUATION'

1. At least two senior R &AW officers—one of them investigated a part of the Purulia arms drop case—were convinced that Bleach sometimes did freelance work for British intelligence. Interpol also had reasons to believe that Bleach did 'odd jobs' for MI6. Annie Machon, a former MI5 officer, has claimed that Bleach was an MI6 agent. See Annie Machon, *Spies, Lies and Whistleblowers: MI5, MI6 and the Shayler Affair*, Sussex: The Book Guild, 2005, p. 387. Machon's conclusion is, however, based on insufficient evidence and largely drawn from newspaper accounts of the Purulia arms drop case.

2. Towards the end of 1995, Bleach, who was in some financial difficulty around that time, successfully secured three defence export contracts from the Bangladesh government. His agent in Bangladesh was one Captain (retired) Syed Tasleem Hussain, who owned a company called Riverland Agencies Ltd, Dhaka, which is accredited to the Bangladesh Ministry of Defence. All of Bleach's tenders used to be submitted through Hussain's company. His other contact in Bangladesh was Iqbal Ali, owner of a freight company in Dhaka. Ali was the key person in the aborted Cargo Air Transport Services (CATS) venture in which Davy and Haestrup were sought to be made partners. Ali was the local agent for an American freight company.

3. Christopher Andrew, *The Defence of the Realm: The Authorised History of MI5*, London, New York, New Delhi: Penguin Books, 2010, p. 788.

4. On 18 August 1995, Bleach had just returned from the day-long meeting in Copenhagen. And yet, a copy of the faxed letter to Allkins showed that he sent out the fax the same day, at least three

days before he spoke to Stuart Mills and then Allkins, which would be 21 August 1995. Did he predate the fax message on purpose?

5. Letter from the UK Organised and International Crime Directorate, Home Office, to Simon Scadden Esq, British Deputy High Commissioner, 1, Ho Chi Minh Sarani, Calcutta 700071, India, 10 June 1998, p. 1.

6. Ibid.

7. Bleach, 'Delivery of Arms by Air to West Bengal', p. 16.

8. DNA was possibly a short form of Davy's Ananda Marga spiritual name, Dada Anindyananda.

9. A suspected Ananda Marga member active in the US, whose identity both the CBI and Interpol are still clueless about.

10. Brannerud, 'Project Purulia', p. 8. Both Interpol and the CBI agree that 'AN' is an abbreviation of Ananda Nagar, the Ananda Marga headquarters in Purulia.

11. Ibid., p. 9. The fax message was typed out on Nielsen's laptop, which was later seized from the AN-26 aircraft after a thorough search at Mumbai airport. The CBI could successfully 'break into' the laptop and cull out almost all details related to the Purulia operation stored in it.

12. Bleach, 'Delivery of Arms by Air to West Bengal', p. 17.

13. Brannerud, 'Project Purulia', p. 9.

14. Witness Statement of Detective Sergeant Stephen Leslie Elcock, 25 November 1996, p. 1. This statement was produced as evidence by the prosecution as well as Bleach during the trial in Calcutta.

15. Marked 'Exhibit 218'. Trial court documents which the prosecution side presented in the course of Sessions Trial No. 1 of June 1997 arising out of CBI/SPE/SIB/Cal/RC 11/S/95 and Jhalda police station Case No. 152 of 1995, dated 18 December 1995. The case related to the *State vs Peter Bleach, Aleander Klichine, Igor Moskvitine, Oleg Gaidach, Euvgeni Antimenko, Igor Timmerman and Vinay Kumar Singh.*

16. Brannerud, 'Project Purulia', p. 9

17. Ibid.

18. Witness statement of Sergeant Elcock, p. 2.

19. Ibid.

20. Ibid.

21. Investigations led the CBI and the IB to describe Randy as an Ananda Marga member with the spiritual name of Suranjanananda

Avadhoot. His real name is presumed to be either Satyendra Narayan Singh and/or Satyanarayana Gowda. He absconded after his name emerged in the course of the investigation.

22. Bleach, 'Delivery of Arms by Air to West Bengal', p. 2 and 18.
23. Ibid., p. 19.
24. Peter Haestrup supposedly told Bleach later that it was their custom to pay the Hells Angels, a bikers' gang, to deal with people who tried to cheat them over money.

5. 'THE RIGHT AIRCRAFT HAS BEEN FOUND'

1. Brannerud, 'Project Purulia', p. 10.
2. Ibid., p. 47.
3. Ibid., p. 10.
4. Cross-examination report of Brian Thune and Peter Haestrup by the Copenhagen Criminal Investigation Department Police on 26 October 1996, p. 27. William Johnson could be identical to the person who Nielsen introduced as 'Bryan' to Bleach at the 27 September 1995, meeting in Bangkok.
5. Ibid., p. 7.
6. Brannerud, 'Project Purulia', p. 11. The R&AW's inquiries indicated that in 1993, under the umbrella of a company called the Northern Scaffold Group PLC, Rosenbloom; a fellow, Englishman Colin Butt; and a Belgian, Edward Stephen Bastian, set up a business importing Russian aircraft to the UK. Out of rivalry, Bastian, who had contacts in Latvia and Russia, was reported to the British authorities by Rosenbloom who complained that Bastian was living in England in breach of his immigration status. This led to Bastian's deportation in July 1994. But by August the same year, Bastian set up a company called Sun City Aero in Riga and along with some Russian and Latvian partners, got involved in selling fighter aircraft and helicopters.
7. Noel Dawes was most likely a retired British Army lieutenant colonel who later became a naturalized American citizen.
8. Witness statement of Mikhail Bruks given to Latvian Interior Ministry security inspector, I. Ulmanis, in Riga on 13 December 1996.
9. Brannerud, 'Project Purulia', p. 11.

10. Bleach, 'Delivery of Arms by Air to West Bengal' p. 25.

11. Brannerud, 'Project Purulia', p. 11. Interpol officers culled this information from a fax message Bleach had sent to Viktor Masagutov, dated 13 October 1995.

12. On 27 October 1995, via another fax message, Bleach had sought Masagutov's help to have Wai Hong Mak 'get through passport control' at Riga airport.

13. Minutes of the meeting of the board of directors of the company held at Suites 1-3, 16th Floor, Kinwick Centre, 32, Hollywood Road, Central, Hong Kong, 11 October 1995. The meeting was chaired by Ms Cecilia D'Cunha, who represented Sovereign Managers Limited.

14. The company, registered in the Isle of Man on 23 May 1988, was dissolved on 17 June 1997.

15. Minutes of the meeting of the board of directors of the company held at Suites 1-3, 16th Floor, Kinwick Centre, 32, Hollywood Road, Central, Hong Kong, 10 November 1995. The chairperson of the meeting was one Sukhvir Kaur Jarhia, a Hong Kong resident of Indian origin whom the CBI tried to track down unsuccessfully. Jarhia represented Sovereign Managers Limited and was responsible for setting up Carol Air Services Ltd and its subsequent administrative work, but was not involved in the actual operation/business of the company. She subsequently emigrated to the US. The other person who attended the meeting was Ms Dolores Butterworth, representing Homeric (TCI) Limited.

16. Bleach, 'Delivery of Arms by Air to West Bengal' p. 32.

17. Baseops Europe is a London-based flight facilitating company, located near Gatwick Airport, which operates worldwide, or at least did so at that time.

18. Peter Bleach's fax message to Baseops Europe, 6 November 1995, p. 2

19. Witness Statement of Aivars Rittenberg, made before Latvian Republic Interior Ministry Security Inspector I. Ulmanis on 25 January 1996. According to Mara Bekere, whose statement was recorded by Ulmanis on 26 January 1996, Nielsen promised her a job at JLY Enterprises Corp. on a montly salary of $500, besides a certain percentage of the commercial transactions, if she signed the contract for the purchase of the AN-26.

20. Certificate issued by Andris Zalmanis, director general, Civil Aviation

Administration, Ministry of Transport of the Republic of Latvia, 21 November 1995.

21. Witness Statement of Viktor Masagutov made before the Latvian Interior Ministry Security Inspector I. Ulmanis on 1 February 1996.

22. Bleach, 'Delivery of Arms by Air to West Bengal' p. 28.

23. Report of the Bulgarian National Investigation Service, prepared and signed by Deputy Director A. Andreev, 19 December 1996.

24. Ibid., p. 29.

25. Arsenal Ltd continues to be one of the largest producers of SALWs in the world, including pistols, sub-machine guns, assault rifles, grenade launchers and mortars, cartridge-based ammunition and rifle grenades. See the Small Arms Survey website at http://www.smallarmssurvey.org/weapons-and-markets/producers/industrial-production.html.

26. Interrogation report of witness Russy Houbanov Russev, executive director, KAS Engineering Co., 10 December 1996, Sofia. Russev was interrogated in the presence of a magistrate at the Bulgarian National Investigation Service, following a Letters Rogatory issued by the chief metropolitan magistrate, Calcutta, on 18 July 1996.

27. Chandan Nandy, 'Bangla Army officer retired for link in arms drop case', *The Telegraph*, Calcutta, 17 March 1998, p. 9.

28. R&AW UO Note of 5 August 1996.

29. Letter of Major M. Jasim Uddin for the director general of the Bangladesh Directorate General Defence Purchase, 25 November 1995. In a subsequent letter of 15 February 1996, the Bangladesh DGDP blacklisted Bleach's Aeroserve UK and Syed Tasleem Hussain's Dhaka-based Riverland Agencies for their 'involvement' in the Purulia arms drop was viewed by the DGDP as an instance of 'direct help to terrorists'. Claiming that Bleach's active participation in the arms drop and the illegal use of the EUC had marred the image of Bangladesh globally, the DGDP moved to cancel all his pending contracts, including one with the Chinese arms manufacturing parastatal NORINCO.

30. Brannerud, 'Project Purulia', p. 14.

31. Ibid., p. 16.

32. Ibid., p. 18.

6. A STRUGGLE, BUT CARGO AWAY

1. Brannerud, 'Project Purulia', p. 16.
2. R&AW UO note of 23 August 1997.
3. R&AW note of 28 December 1995.
4. Civil Aviation Authority of Bangladesh Assistant Director Mohammad Habibullah's letter to Air Parabat Ltd, dated 29 November 1995.
5. It was later discovered that 300 Kalashnikov rifles and 25,000 rounds of ammunition, besides rocket-propelled grenades, several grenade launchers, Makrarov 9-mm pistols and some Dragunov rifles were finally packed and loaded onto the AN-26 aircraft. Bleach revealed to the CBI that he was not aware how weapons other than the AK-47 rifles were included in the consignment, especially when the original order was only for Kalashnikov rifles and matching ammunition.
6. A most curious and mysterious event came to the notice of Interpol on 11 December 1995. That day a money transaction against Nielsen's Visa account was recorded. A payment of $103 was made to 'Wheels Rent-a-Car' in New Delhi. Other records and documents show that Nielsen was at Bourgas Airport the same day. Could it then be that someone else was operating his Citibank Visa card while he was in Bourgas? This is what most probably happened after it was discovered that the particular Citibank Visa card was being used by its primary holder, Peter Brook. At the same time, neither Interpol nor CBI could determine with certainty whether a William Johnson (Brook's alias) stayed at Hotel Sofitel in Bourgas when the arms and ammunition were being loaded onto the aircraft. A William Johnson had previously checked in with Nielsen at Calcutta's Great Eastern Hotel on 18 September 1995. In all likelihood, William Johnson, whoever he was, accompanied Nielsen to Purulia to 'inspect' whether an AN-26 could land on a rough strip around Purulia to deliver the weapons directly to the insurgents.
7. The part of an airport nearest to the aircraft, the boundary of which is the security check, Customs and passport control.
8. Brannerud, 'Project Purulia' p. 19.
9. Ibid.
10. Bleach, 'Delivery of Arms by Air to West Bengal', p. 32.

11. According to Brannerud's Project Purulia document (p. 17), Bleach had said that when he wanted to return home after the arms and ammunition were loaded onto the aircraft, Nielsen made a veiled threat. Bleach's wish to return to the UK was 'backed up' by statements that 'Bleach wouldn't want anything to happen to his twenty-three-year-old daughter or to his elderly mother.'

12. Interpol Red Corner Notice No. A/177/4/1996. Deepak was later identified as Deepak Manikan, alias Daya M. Anand, born on 24 March 1961. He is suspected to be an Ananda Marga member based at that time in Singapore. His Singapore passport number was S-1477707 issued in Singapore on 7 April 1993.

13. Bleach, 'Delivery of Arms by Air to West Bengal', p. 33.

14. Bleach, 'Delivery of Arms by Air to West Bengal', p. 33.

15. Ibid, p. 34.

16. Ibid.

17. Brannerud, 'Project Purulia', p. 21.

18. Bleach, 'Delivery of Arms by Air to West Bengal', p. 35.

19. Indian Air Force headquarters' letter to the Ministry of Defence on 31 December 1995.

20. Bleach, 'Delivery of Arms by Air to West Bengal', p. 36.

21. Indian Air Force report of Group Captain C. de Souza, deputy director (Directorate of Air Intelligence), 20 January 1997.

7. NO RECEPTION COMMITTEE AT MUMBAI AIRPORT

1. Author's interview with Pranab Kumar Mitra at Barasat near Calcutta, on 12 October 2012.

2. Purulia District Gazette, West Bengal government, 1985, p. 108.

3. Jhalda Police Station Case Diary under Section 172 of the Criminal Procedure Code. Mitra's notes in the Case Diary spanned over a period of three days—18-20 December—before it was handed over to the West Bengal Criminal Investigation Department (CID).

4. Ibid.

5. Purulia District Gazette, p. 1.

6. Bleach, 'Delivery of Arms by Air to West Bengal', p. 40.

7. Ibid.

8. An apron side is usually the area where aircraft are parked and unloaded or loaded, refuelled or boarded.

9. Proceedings of the Anti-Terrorist Law Group Meeting held in the Maharashtra chief secretary's committee room on 18 January 1996, p. 5. The meeting was presided over by the additional chief secretary (home).

10. Ahamed et al., 'Third Report Purulia Arms Dropping', p. 19.

11. Hong Kong police authorities later shared Wai Hong Mak's phone number, which matched with the one the call was made to from Mumbai airport on 22 December 1995. Investigations later revealed that Nielsen had called up the same number during a previous visit to Mumbai in 1994, when he stayed at the Ritz Continental Hotel in Churchgate.

12. Chandan Nandy, 'Davy called arms drop financier from Mumbai airport', *The Telegraph*, 24 January 1996, p. 1.

13. Witness statement of Vladimir Ivanov made before Latvian Interior Ministry Security Inspector I. Ulmanis on 2 February 1996.

8. 'YOU CONSPIRED TO WAGE WAR'

1. The Ananda Marga's registered headquarters is located at Ananda Nagar in West Bengal's Purulia district, but it actually functions from a large location in Tiljala, on Calcutta's eastern fringes. The organization commenced functioning outside in the 1960s and had nine sectors with several sub-regions. Those sectors were: New York, Berlin, Hong Kong, Manila, Sydney, George Town (Guyana), Cairo, Nairobi, and Delhi. Every Ananda Margi uses a Sanskrit name given to him/her by the organization and many of them have used such names to conceal their real identity. The Ananda Marga philosophy has two facets: religious philosophy and political philosophy. A small section within the Ananda Marga believed in and practised violence. The founder, P.R. Sarkar, did not tolerate any dissent within the organization and deviants and defectors who spoke against the order were purged from time to time. Sarkar and some of his followers were arrested in 1971 for their alleged complicity in the murder of eighteen Margis who chose to defect and dissent. Following an imposition of a ban on the Ananda Marga in July 1975 by the Indian government, the organization took to agitations and violence in India and abroad. After his trial, Sarkar and some of his associates were found guilty of murder and sentenced to life

imprisonment. Following an appeal against the conviction, a high court released Sarkar in August 1978. The organization fragmented after 1990 when Sarkar passed away. There are as many as three factions, each claiming primacy over the other, who control Sarkar's legacy or whatever is left of it and an intense power struggle continues among the groups. See Stephen A. Kent, *From Slogans to Mantras: Social Protest and Religious Conversion in the Late Vietnam War Era*, New York: Syracuse University Press, 2001, p. 47. See also Helen Crovetto, 'Ananda Marg, Prout, and the Use of Force', in James R. Lewis (ed.) *Violence and New Religious Movements*, New York:Oxford University Press, 2011. p. 456.

2. Author's interview with Pranab Kumar Mitra at Barasat near Calcutta on 12 October 2012.

3. Ministry of Home Affairs Suspect Index, 1996.

4. Chandan Nandy, 'Indian mole may have fled to Colombo', *The Telegraph*, Calcutta, 31 December 1995, p. 1. See also by Chandan Nandy, 'Davy top gold smuggler: CBI', *The Telegraph*, Calcutta, 1 January 1996, p. 1.

5. Witness statement of Zainulabidin Abdul Karim Surve to the CBI in Mumbai on 30 December 1995.

6. Witness Statement of Amirali Chagan Bawa to the CBI in Mumbai on 30 December 1995.

7. Ibid.

8. Witness Statement of R. Jyoti Prasad to the CBI in Mumbai on 2 January 1996.

9. Ibid.

10. Director, Intelligence Bureau, D.C. Pathak's response on 8 July 1998 to the CBI's questionnaire of 19 June 1998, p. 1.

11. Ibid., p. 2

12. The CBI's 'confidential', unsigned and undated note on 'Lapses on the Part of Government Agencies/Senior Officers in Purulia Arms Drop Case', pp. 7–10.

13. R&AW Special Bureau Memorandum No. V/SB/BOM/VII/94(17) of 22 January 1996, p. 1.

14. Minutes of the meeting of the Maharashtra Anti-Terrorist Group chaired by the Maharashtra additional chief secretary (home), State Secretariat, Mumbai, on 18 January 1996, p. 5.

15. Ahamed et al., 'Third Report Purulia Arms Dropping', p. 15.

16. Statement of Secretary to the Ministry of Civil Aviation Yogesh Chandra to the Parliamentary Committee on Government Assurance, on 19 March 1997, in Ahamed et al., 'Third Report Purulia Arms Dropping', p. 15.

17. Bleach, 'Delivery of Arms by Air to West Bengal', p. 39.

18. Witness Statement of T. Tamil Selvan recorded by CBI Inspector M. Muruganandam in Chennai on 25 January 1996.

19. Ibid.

20. Witness Statement of S.G. Sathe recorded by CBI Inspector N. Muruganandam on 25 January 1996, p. 1.

21. Ibid, p. 2.

22. Ibid.

23. Ahamed et al., 'Third Report Purulia Arms Dropping', p. 28.

24. Witness Statement of Assistant Commissioner of Customs K. Umesh recorded by CBI Inspector N. Muruganandam in Chennai on 19 January 1996, p. 3.

25. Witness Statement of Chennai Airport Security Inspector V. Pattaviraman, recorded by CBI Inspector N. Muruganandam in Madras on 10 January 1996.

26. Report of the Inter-Departmental Meeting on Granting Flight Clearance to Non-Scheduled Flights, 14 March 1997, p. 3.

27. Witness Statement of Customs Superintendent Om Prakash, recorded by CBI Deputy Superintendent of Police (Lucknow) Kuldip Singh in Varanasi on 1 January 1996, p. 2.

28. Report of the Inter-Departmental Meeting on Granting Flight Clearance to Non-Scheduled Flights, 14 March 1997, p. 1.

29. A term used in West Bengal, Assam and different parts of Bangladesh to refer to a small hamlet or village. In the nineteenth and twentieth centuries, the term referred to a revenue collection unit in a district. Of late, however, mouza has declined in usage and importance.

30. Witness statement of Kartik Pramanik, recorded by CBI Inspector B.K. Bagchi at Garudih village on 18 January 1996.

31. Witness Statement of Montu Gorai, recorded by CBI Inspector M.K. Sarkar, at Garudih village on 16 January 1996.

32. Witness statement of Barun Chandra Kumar, recorded by CBI Inspector B.K. Bagchi at Baro Rolla village on 18 January 1996.

33. Witness statement of Paritosh Saha, recorded by CBI Inspector B.K. Bagchi at Taherbera village on 18 January 1996.

34. Witness Statement of Riju Majhi, recorded by CBI Deputy Superintendent of Police A.N. Moitra, in Bhagudih village on 14 July 2004, three months after the arrest of Tadbhavananda (who was seventy-three-years-old at that time) from Malviya Nagar, Delhi.

35. Chandan Nandy, 'Arms drop site under illegal Marg occupation' *The Telegraph*, Calcutta, 11 May 1997, p. 7.

36. Report issued by the Office of the District Land and Land Reforms Officer, Purulia, addressed to CBI Special Crime Branch Inspector A.K. Das, on 8 December 1996.

37. CBI chargesheet of 20 March 1996, pp. 6–7.

38. Analysis of GPS data by IAF Group Captain T.R. Rau, communicated to the CBI vide Secret Air Headquarter Note No. 20192/6/AI(S) PC-8, 13 February 1996, p. 2.

39. Analysis of GPS data by IAF Group Captain T.R. Rau, communicated to the CBI vide Secret Air Headquarter Note No. 20192/6/AI(S) PC-8, 13 February 1996, p. 7.

40. Survey of India Secret Report No. 63/44P of 29 February 1996, p. 2. The results were achieved after the waypoints were plotted on topographical maps.

41. Confidential Air Headquarters report prepared by Group Captain C. de Souza, 20 January 1997.

42. CBI chargesheet of 20 March 1996.

43. Ibid.

44. Ibid.

45. After his arrest in April 2004, Tadbhavananda passed away because of cancer.

46. CBI chargesheet against Acharya Tadbhavananda Avadhoot, alias Lal Chand Parihar, alias Tushar Kanti, originally from Doda district in Jammu, on 16 November 2004, p. 6.

47. Ibid., p. 9.

48. Author's interview with Pranab Kumar Mitra at Barasat near Calcutta on 10 October 2012.

49. Trial court judgement, arising out of CBI's case RC/11/S/95 and Jhalda police station Case No. 152 of 1995, 31 January 2000, p. 34.

50. Ibid., p. 60.

51. Jon Stock, 'Blast from Bleach: Exclusive Letter from Jail', *The Week*, 20 July 1997, p. 14.

52. Ibid, pp. 15 and 17.

53. Author's telephonic interview with Jayant Umranikar on 5 December 2012.

54. Brannerud, 'Project Purulia', p. 35.

55. Brannerud, 'Project Purulia', p. 36.

56. Chattopadhayay, Suhrid Shankar, 'Waiting to Go Home', *Frontline*, Vol. 18, Issue No. 10, 18-25 May 2001. Accessed from http://www.frontlineonnet.com/fl1810/18100420.htm, on 14 October 2012.

57. Taylor, Sir Teddy, Speech on Peter Bleach in the House of Commons, 27 November 2002. Accessed from the website of the British Parliament at http://www.publications.parliament.uk/pa/cm200203/cmhansrd/vo021127/halltext/21127h03.htm, on 14 October 2012

58. Ibid.

59. Graham Brough, 'Riddle of the Tory MP, the Mercenary and Indian Jailbreak: Plot to Free Arms Dealer Pal', *The Mirror*, 22 September 1997.

60. Ibid.

61. Author's telephonic interview with Christopher Hudson on 28 December 2012.

62. Sir Richard Dearlove's email to author on 8 January 2013.

63. Author's telephonic interview with Hudson on 28 December 2012.

64. Calcutta High Court judgement of Justice A.K. Ganguly, 22 September 2002. Accessed from http://indiankanoon.org/doc/1898988/?type=print on 14 October 2012.

65. Ibid.

66. Unsigned and undated MHA note expressing the ministry's arguments and the grounds for not releasing Bleach, p. 2.

67. Ibid, pp. 6–7.

68. Ibid, p. 8.

9. THOU SHALT NOT GET CAUGHT

1. Chandan Nandy, 'Davy hid in Osho ashram after arms drop', *The Telegraph*, Calcutta, 11 March 1996, p. 7. The CBI was able to track down the taxi driver who admitted that he was woken up from sleep by a 'white man' who later tipped him generously.

2. When I checked with a classified Indian intelligence document on a number of 'blacklisted' Ananda Margis, I was most surprised to come across Steven Michael Dwyer's name. Every detail about Dwyer matched with the findings of the R&AW, including his exact role in the stabbing of the Indian Embassy staff in Manila on 7 February 1978, and the others who had taken part in the fatal assault.

3. Nielsen's Hong Kong address was 22-36, Patterson Street, 21 Floor, C-3, Pearl City Mansion, Causeway Bay, Hong Kong.

4. Brannerud, 'Project Purulia', p. 23.

5. R&AW UO Note of 10 December 1996.

6. Brannerud, 'Project Purulia', p. 64.

7. Ibid.

8. Ibid.

9. Ibid.

10. Ibid.

11. Ibid. In the early and mid-nineties, Sudan was the most-favoured destination for international terrorists, including Osama bin Laden and Ilich Ramirez Sanchez, better known as 'Carlos the Jackal' who was captured in Khartoum in the first week of August 1994 and sent to France to stand trial for the assassination of two French counter-intelligence officers and a Lebanese national.

12. Sometime after he was brought to Calcutta following his arrest in Mumbai, Joel Broeren performed tandava nritya (dance of destruction) at the Calcutta branch office of the CBI. Among other things, it proved that he was indeed an Ananda Margi.

13. Interpol Red Corner Notice No. 36650/2004 against Martin Konrad Schneider, http://www.cbi.nic.in/rnotice/38.A-1213-8-2004.html

14. Author's interview with Joel Broeren in Calcutta after he was released on bail, on 10 July 1997.

15. Joel Broeren's statement to the CBI, p. 7.

16. Ibid.

17. Chandan Nandy, 'Armsdrop aide flies the coop', *The Telegraph*, Calcutta, 17. January 1999, p. 1.

18. Bleach, 'Delivery of Arms by Air to West Bengal', p. 45.

19. Ibid.

20. Stig Toft Mardsen, 'A Dane in distress', blog posted on AsiaPortal, 3

October 2011. Accessed from http://infocus.asiaportal.info/2011/
10/03/a-dane-in-distress/, on 17 October 2012.
21. Howerstoke Trading Ltd was dissolved on 15 May 1998.
22. Undated fax message from Niels Christian Nielsen, alias Peter
Johnson, to Ivan Whitehead in Johannesburg, sent from a Hong
Kong fax number, 852-2882-1723.
23. Nielsen paid $7,000 for the false passport in the name of the dead
four-week-old infant, Kim P. Davy. In all likelihood it was procured
for him by the suspected Irish Republican Army member Perry
'Mick' Joseph.
24. R&AW UO Note of 6 January 1997.
25. Interpol Red Corner Notice, Control No. A-652/12-1997.
26. In an article in the Danish newspaper *Politiken* (7 September 2007),
Nielsen was quoted as claiming that Prabhat Ranjan Sarkar, the
founder of Ananda Marga, was a nephew of Subhas Chandra Bose.
Accessed from http://politiken.dk/debat/kroniker/ECE375007/de-
kalder-mig-terrorist/ on 18 October 2012.
27. Kaare Skovmand, 'Portrait: Naive adventurer or terrorist?' *Politiken*,
Copenhagen, 9 April 2010. Accessed from http://politiken.dk/
indland/ECE943517/portraet-naiv-eventyrer-eller-terrorist/ on 17
October 2012.
28. Cross-examination report of Brian Thune and Peter Haestrup by
the Copenhagen Criminal Investigation Department Police on 25
October 1996, p. 40.
29. Danish Ministry of Justice's letter of 25 May 2000 to the Indian
Embassy, Copenhagen, p. 2.
30. Ibid.
31. Danish Ministry of Justice's letter of 22 Janaury 2001, to the Indian
Embassy, Copenhagen, p. 1.
32. Ibid, p. 2
33. Quoted in the judgement of the Hillerod Court in case SS9-1316/
2010 the Danish Ministry of Justice brought against Niels Holck
under civil registration number 171261-1061, on 1 November 2010,
p. 27.
34. Brannerud, 'Project Purulia', p. 41.
35. Author's telephonic interview with William Roeschke on 29
December 2012.
36. Judgement of the Hillerod Court in case SS9-1316/2010 the

prosecution brought against Niels Holck under civil registration number 171261-1061, on 1 November 2010, p. 2.

37. Ibid., p. 14.
38. Ibid., p. 17.
39. Ibid., p. 13.
40. Ibid., pp. 8-9.
41. Ibid., pp. 11-12.
42. Ibid., p. 12.
43. HT Correspondent, 'CBI seeks Davy extradition, with expired warrant', *Hindustan Times*, New Delhi, 19 May 2011, p. 1. No action was taken against senior CBI officers for this lapse. Only an inspector was suspended.
44. Judgement of the Eastern High Court's 11th Department of J.Nr. S-3321-10 of 6 June 2011. The Danish High Court's decision in respect of Nielsen was in stark contrast to the one related to Camilla Broe, the first Danish national extradited to the United States (outside the European Union) on charges of drug trafficking, in September 2009, the same year the two countries signed an extradition treaty. Broe, who was accused by the US Drug Enforcement Agency for smuggling in several thousand pills of Ecstasy to Florida, was extradited without any fuss, though it later turned out that her case was barred by the statute of limitation.
45. Author's conversation with CBI Deputy Inspector General Arun Bothra on 20 November 2012.
46. Author's conversation with CBI Deputy Inspector General Arun Bothra on 20 November 2012.
47. 'Debate: The truth about Purulia-1-4', *Times Now*, broadcast on 28 April 2011.
48. Jorgen Steen Sorensen's statement of 7 July 2011 accessed from the Danish Ministry of Justice's website at http://www.rigsadvokaten.dk/page73.aspx?recordid73=337 on 25 October 2012.
49. 'International Law Commission: The Obligation to Extradite or Prosecute (*Aut Dedere Aut Judicare*)', *Amnesty International Publication*, 2009, p. 8-9. The specific international conventions and treaties to which Denmark is a signatory or has ratified, and to which the 'aut dedere aut judicare' obligation applies, include the International Convention for the Suppression of Terrorist Bombings, 1997, International Convention for the Suppression of Financing of

Terrorism, 1999, United Nations Convention against Transnational Organised Crime, 2000, and the Protocol Against the Illicit Manufacturing of and Trafficking in Firearms, Their Parts and Components and Ammunition, 2001, Supplementing the UN Convention Against Transnational Organised Crime, 2001.

50. Amitav Ranjan, 'To get Kim Davy extradited, Delhi puts Denmark ties in deep freeze', *The Indian Express*, New Delhi, 17 August 2011, p. 1.

51. Feinstein, *The Shadow World*, p. 155.

10. A TRUE MANCHURIAN CANDIDATE

1. CBI Note on Financial Aspects of the Purulia Arms Air-Dropping Case, 1 January 1997, p. 10.

2. Ibid.

3. Peter Bleach's bail petition to the Calcutta High Court, 8 July 1997, p. 14.

4. Ibid.

5. K.S. Subramanian, *Political Violence and the Police in India*, New Delhi: Sage Publications, 2007, p. 138.

6. Anirban Roy, *Prachanda: The Unknown Revolutionary*, Kathmandu: Mandala Book Point, 2008, p. 51.

7. Niels Holck and Øjvind Kyrø, *De Kalder Mig Terrorist* (They Call me a Terrorist), People's Press: Copenhagen, 2008, p. 13.

8. CBI Note on Financial Aspects of the Purulia Arms Air-Dropping Case, 1 January 1997, p. 9.

9. Ibid.

10. Bertil Lintner, *Great Game East: India, China and the struggle for Asia's most volatile frontier*, New Delhi:HarperCollins, 2012, pp. 344–345.

11. Bleach, 'Delivery of Arms by Air to West Bengal', p. 47.

12. Report of the Hong Kong police's Organised Crime and Triad Bureau, 3 December 1997, p. 4. The report was prepared after Mak was questioned by the Hong Kong police.

13. Ibid.

14. Gist of Wai Hong Mak's statement to the Hong Kong Police, 3 December 1997.

15. Brannerud, 'Project Purulia', p. 40.

16. Ibid.

17. CBI dossier on Niels Christian Nielsen, July 2004, p. 3.

18. Brannerud, 'Project Purulia' p. 17.

19. Witness statement of Viktor Masagutov, director of Eastern European Aviation Consultants Ltd, recorded by Riga-based Interpol Chief Specialist Viktor Tuskevich, on 2 January 1997. According to Masagutov, the spare parts were worth $25,000 which Nielsen paid for in cash.

20. Bleach, 'Delivery of Arms by Air to West Bengal', p. 30.

21. Douglas Farah, and Stephen Braun, *Merchant of Death: Money, Guns, Planes, and the Man Who Makes War Possible*, John Wiley & Sons: New Jersey, 2008, p. 320.

22. Jon Mason-Ponting, 'Project Operation Bloodstone: Analytical Report' Criminal Analysis Branch Directorate, Interpol General Secretariat, Lyon, 4 December 2000, p. 12. See also Victor But: Transnational Criminal Activities, Intelligence Brief, US Department of Treasury Bureau of Alcohol, Tobacco and Firearms, December 2000, p. 31.

23. Final Report of the United Nations Monitoring Mechanism on Angola Sanctions, UN Security Council Committee, 21 December 2000, p. 55.

24. Ibid., p. 15. According to an October 2001 report of the UN Monitoring Mechanism on Angola, Samuel Sieve was the sole director of East European Shipping Corporation in the nineties. However, all activity of the corporation ceased in May 2000. On the other hand, Trade Investment International Ltd ceased trading in February 2000. Sieve's other company, Trade Investment UK Ltd, located at Cumberland Mansions, George Street, London, remained active in the UK.

25. Bleach, 'Delivery of Arms by Air to West Bengal', p. 28.

26. UN Report on Angola, 21 December 2000, pp. 38–53.

27. Fax message to 'Kim Davy' from Capt. Ahmed A. Hussain, chief of flight operations of Air Parabat Limited, on 12 December 1995.

28. Intelligence Brief, US Department of Treasury Bureau of Alcohol, Tobacco and Firearms, pp. 10–11.

29. 'Out of Control: The loopholes in UK controls in the arms trade' Oxfam, December 1998, p. 6.

30. UN Report on Angola, p. 32.

31. Feinstein, *The Shadow World*, p. 505.

32. Ibid.

33. Ministry of Home Affairs Note of 5 January 1996.

34. Author's telephonic interview with Jayant Umranikar on 5 December 2012.

35. North Yorkshire police man Detective Constable Jones' handwritten notes, based on the replies given by Bleach to questions posed to him by Detective Sergeant Stephen Leslie Elcock.

36. Letter of G.W. MacAlister to British deputy high commissioner in Calcutta, Simon Scadden, 10 June 1998, p. 1.

37. Ibid., p. 2.

38. Ibid.

39. Brannerud, 'Project Purulia', p. 18.

40. Cabinet Secretariat statement issued by the Press Information Bureau of India in New Delhi on 30 April 2011.

41. Author's telephonic conversation with Ranjita Ranjan on 3 January 2013. Ranjita Ranjan is a former MP representing Saharsa constituency. She joined the Congress in April 2009.

42. Rajesh Ranjan, or Pappu Yadav as he was popularly known, was a four-time MP representing the Rashtriya Janata Dal, the party run by former Bihar Chief Minister Lalu Prasad Yadav, from Purnea. In 2008 he was sentenced to life imprisonment for the murder of a CPI (M) leader Ajit Sarkar in Purnea, Bihar, and is now serving time in Delhi's Tihar Jail.

43. Vir Sanghvi, 'Parallax view: So are we any wiser about the Purulia arms drop?' Accessed from Sanghvi's blog http://www.virsanghvi.com/Article-Details.aspx?key=629 , on 29 October 2012.

44. Holck and Kyro, De Kalder Mig Terrorist, p. 4.

45. Author's telephonic interview with Jayant Umranikar, who was a career intelligence officer before repatriating to his parent Indian Police Service cadre of Maharashtra in 2004, on 5 December 2012.

46. Author's telephonic interview with Bahukutumbi Raman on 28 December 2012.

47. For a detailed account of the CIA's clandestine struggle in Tibet, see Kenneth Conboy and James Morrison, The CIA's Secret War in Tibet, Lawrence: University Press of Kansas, 2002, p. 301.

48. Bertil Lintner, Great Game East: India, China and the struggle for Asia's most volatile frontier, New Delhi: HarperCollins, 2012, p. xvii

49. Bleach, 'Delivery of Arms by Air to West Bengal', pp. 48--49. Bleach was apparently referring to the CIA's covert operations to supply weapons, by using European arms dealers and gunrunners, to the Bosnian Muslims in the early nineties. Since the arms drop operation involved two countries—Bulgaria and Latvia—which continued to be in the zone of Moscow's influence even after the end of the Cold War, the initial suspicion was that the successor to the KGB's First Chief Directorate, the Sluzhba Vneshney Razvedki (SVR) or the Foreign Intelligence Service, was behind the operation. At the time, however, the SVR was more focused on the small republics like Chechnya where an extremist Islamist movement, that would soon assume a violent character, was on the rise. Some R&AW officers, believing Wai Hong Mak's close involvement and Hong Kong as the base of operations, suspected the hand of Chinese intelligence. The objective, according to one senior retired R&AW operative, was to arm the ultra-left movement that was 'picking up in the mid-nineties and such a heavy dose of weapons would have given them the requisite boost.'

50. Author's telephonic interview with Peter Bleach on 28 December 2012.

51. Author's telephonic interview with Lou Girodo who was sheriff of Las Animas between 1988 and 2000.

52. IMMARSAT is a British satellite telecommunications company which offers global mobile services. It provides telephone and data services to users worldwide, through portable or mobile terminals which communicate to ground stations via geo-stationary telecommunications satellites.

53. Clayton Thyne, 'Sudan (1983-2005)', in Karl DeRouen and U.K. Heo (eds), *Civil Wars of the World: Major Conflicts Since World War II*, ABC-CLIO: California, 2007, p. 865. See also Jeffrey Gentleman and Michael R. Gordon, 'Pirates' Catch Exposed Route of Arms in Sudan', *The New York Times*, 8 December 2010. Accessed from http://travel.nytimes.com/2010/12/09/world/africa/09wikileaks-tank.html?n=Top/Reference/Times%20Topics/People/G/Gordon,%20Michael%20R.?ref=michaelrgord on 17 October 2012. See also Horand Knaup, 'Hijacked Weapons: A Discreet Deal for the War in Sudan', *Der Spiegel Online*, 12 September 2010. Accessed from http://www.spiegel.de/international/world/hijacked-weapons-

a-discreet-deal-for-the-war-in-sudan-a-733775.html on 20 October 2012.

54. Peter Wright with Paul Greengrass, *Spycatcher*, Victoria: Heinemann, p. 260.

55. Brannerud, Project Purulia, p. 21.

56. Altagem Resources Inc's letter to SPLA Chairman and Commander-in-Chief Dr John Garang de-Mabior, 25 September 1995. Parts of the printed text on the letter have faded, making it difficult to read the entire contents. The letter was also faxed by an SPLA leader, Samuel T., to one Ivan Whitehead in Johannesburg, with whom Nielsen had had more than one meeting in the South African capital in 1995.

57. See Jaime C. Zajra's LinkedIn profile at http://ph.linkedin.com/pub/jaime-zafra/52/24/52. Accessed on 25 October 2012.

58. Author's telephonic interview with Jaime Zafra on 25 October 2012.

59. Jaime Zafra's email of 25 October 2012 in response to the author's questions.

60. Christopher Robbins, *Air America: From World War II to Vietnam*, Bangkok: Asia Books, 2003, p. xxi.

61. Sir Edward Taylor, 'Discussion on Peter Bleach, House of Commons Hansard Debate', Part 3, Column 114WH, 27 November 2002. Accessed from http://www.publications.parliament.uk/pa/cm200203/cmhansrd/vo021127/halltext/21127h03.htm on 16 December 2012.

62. Subir Bhaumik, 'Kim Davy in Kenya?' *The Week*, 20 February 2000.

63. Author's telephonic interview with Christopher Hudson on 28 December 2012.

64. Author's telephonic interview with retired US Bureau of Alcohol, Tobacco, Firearms and Explosives Special Agent Donald Manross on 28 November 2012.

65. For reasons of source protection, I cannot reveal the names of either the sender or the receiver of the emails.

66. George Church, 'The US and Iran', *Time*, 17 November 1986.

67. Amnesty International Report, 'Denmark: A Briefing for the Committee Against Torture', 27 April 2007, pp. 9–11. An investigative story by the Danish newspaper *Politiken* on 21 October 2007, claimed that one of the planes known to have been used for

CIA rendition flights was given permission to cross Danish airspace on 25 October 2005. The report revealed that the plane, which was en route from Washington to Jordan, picked up a Yemeni national, Muhammad Bashmillah, from illegal detention in Jordan and from there rendered him to secret US custody.

68. Ibid.

69. Ibid., p. 11. Amnesty International highlighted in particular the case of five men from Odense, Denmark, charged with attempted violation of section of 114 of the Criminal Code (the primary provision criminalising act of terrorism), who were suspected of having purchased chemicals to build explosive devices with a view to committing an act of terrorism at an unknown location in Denmark. They were held in solitary confinement from 5 September 2006 to 16 January 2007, pending the completion of investigation and issuance of the indictment.

70. Author's telephonic interview with Bahukutumbi Raman on 28 December 2012. Raman passed away on 16 June 2013.

71. An arrangement covering hiring of an aircraft, including the provision of a flight crew and sometimes fuel.

72. Champion, 'Subreptitious Aircraft in Transnational Covert Operations', p. 465.

73. Witness statements of Alexander Lukin and Vladimir Ivanov, made before Interpol Chief Specialist Viktor Tuskevich in Riga on 24 and 28 December 1996, respectively.

74. Author's telephonic interview with Donald Manross on 17 November 2012.

75. Author's telephonic interview with Jayant Umranikar on 5 December 2012.

76. Ibid.

77. Author's interview with Ajit Doval on 20 November 2012.

78. Ibid.

79. The administrative capital of Myanmar was officially moved to Naypyidaw, about 320 kilometres north of Yangon, on 6 November, 2005. The capital's official name was announced on 27 March 2006.

80. Peter Bleach's fax advice to Baseops Europe, sent on 6 November 1995. Note that the advice was made fifteen days before the AN-26 aircraft was purchased (on 21 November 1995) from Latavio in Riga, Latvia.

81. Baseops Europe's flight plan for Peter Bleach, 7 December 1995. The three-letter codes are KHI (Karachi), VNS (Varanasi), DAC (Dhaka) and RGN (Rangoon). The times indicated were in accordance with UTC or Coordinated Universal Time, a modern continuation of Greenwich Mean Time. LT stands for Local Time. ETA and ETD is expected time of arrival and departure.

82. Bleach, 'Delivery of Arms by Air to West Bengal', p. 34.

83. Baseops Europe's telex message of 16 December 1995, to Peter Bleach. The telex message was marked for the front desk duty manager at Pearl Continental Hotel, Karachi.

84. Brannerud, 'Project Purulia', p. 61.

85. Bleach, 'Delivery of Arms by Air to West Bengal' p. 37.

86. Item No. 488 on the CBI's List of Documents and Articles collected in relation to the agency's Case No. RC/11/SCB/95-Cal, p. 28.

87. Bleach, 'Delivery of Arms by Air to West Bengal', p. 49. In a lengthy telephonic interview with me on 29 December, Bleach admitted that he had indeed been fed with a lot of disinformation by Nielsen.

88. J.M, 'Myanmar's minorities: Caught in the middle', The Economist, 11 July 2012. Accessed from www.economist.com/blogs/banyan/2012/07/myanmars-minorities, on 30 December 2012.

89. The ceasefire lasted until June 2011, when fierce fighting broke out between the KIA and the Myanmar army.

90. Steve Coll, Ghost Wars: The Secret History of the CIA, Afghanistan and Bin Laden, From the Soviet Invasion to September 10, 2001, London: Penguin Books, 2004, p. 119.

91. Lintner, Great Game East, p. 255.

92. Ibid.

93. Ibid.

94. NORINCO is a Chinese company that enjoys substantial support from the government. It manufactures high-tech defence products, including precision strike systems, amphibious assault weapons and equipment, long-range suppression weapons systems and anti-acraft and anti-missle systems.

95. Subir Bhaumik, 'Kim Davy in Kenya?' The Week, 20 February 2000.

96. Author's telephonic interview with Christopher Hudson on 28 December 2012. When Hudson met Bleach at Presidency Jail, he (Hudson) was the national secretary of the right-wing Conservative Monday Club.

Armsdrop papers ploy to throw sleuths off scent, fear agencies

Cache issued in Bangla officer's name

2 2 MAY 1996

BY CHANDAN NANDY

Calcutta, May 21: The "end user" certificate for Kim Peter Davy's weapons consignment, that was airdropped in Purulia last December, was issued in the name of a senior officer in the Bangladesh Army.

Documents in possession of the investigating agencies probing the sensational armsdrop disclose the name of a certain Major General Subed Ali Bhuyan, Principal Staff Officer of the Armed Forces Division, in Begum Khaleda Zia's Secretariat, as the man who had signed the end user certificate.

An "end user certificate" is a legitimate document authorising a person or organisation to transport arms and ammunition to a consignee.

It is, however, also likely that Kim Peter Davy, and the others involved in the armsdrop case, could also have forged the end user certificates by using fake signatures of Major General Bhuyan and fake seals of the Dhaka government to show, wherever and whenever necessary, that the arms were meant for delivery to the Bangladesh military.

It may be recalled that, soon after the sophisticated consignment of weapons — that included a huge cache of AK-47 and AK-56 assault rifles, 0.9mm pistols, rocket launchers and rocket-propelled grenades — were recovered by the West Bengal Police in Purulia, it was found that the crates containing the arms had the name of "Rajendrapur Cantonment" inscribed on them. Rajendrapur Cantonment is situated in Dhaka district.

There was considerable speculation at that time about the possible end users of the arms.

It was believed that the arms were meant for Bangladesh, and subsequent investigations revealed that the Latvian An-26 aircraft which was used to airdrop the weapons, had been scheduled to land in Dhaka, but was denied permission by the Bangladeshi authorities.

No clues, however, were later found by Indian authorities to link the armsdrop to Bangladesh.

Officials of the Central Bureau of Investigation (CBI), however, did find out from the six arrested An-26 crew members that Kim Peter Davy had intended to get the aircraft, originally purchased from Latvian Airlines, registered in Dhaka.

Interestingly, Peter Bleach, a former member of the British military intelligence, and a part of the arrested An-26 crew, had told CBI interrogators that, according to their original plan, he was to use his "contacts" in the Bangladesh military to get the aircraft registered there.

CBI sources agreed that "a more comprehensive investigation" was required to find out if Major General Bhuyan was at all linked with Kim Peter Davy.

Fearing cases of "unreported" armsdrop or despatch of arms and ammunition over Indian territory in the past, Union home ministry officials said "Major General Bhuyan may have been involved in issuing end user certificates to organisations transporting arms to the National Socialist Council of Nagaland (NSCN), Muivah faction, from South East Asian countries."

In this context, sources said that the Union home ministry had specific information about the routes followed by the NSCN rebels to smuggle arms through Bangladeshi territory.

Over the past five years, six arms consignments of the NSCN had been detected in Bangladesh territory. This year, two arms consignments have already come to the notice of Indian authorities.

The latest instance of persons caught while trying to smuggle arms into India occurred in March, at Cox's Bazar, in Bangladesh's southern coastal district of Chittagong. The cache included AK-47 rifles, grenades and machine guns.

But, sources pointed out, the detection of the arms by the Bangladeshi authorities was purely "by chance," since at that time police officials were conducting raids in several parts of that country on the eve of elections.

Indian mole may have fled to Colombo

3 DEC 1995

BY CHANDAN NANDY

Calcutta, Dec. 30: A preliminary Central Bureau of Investigation (CBI) report has indicated the mysterious Randy, being touted as frontman for the recipient of the Purulia arms consignment, is actually Randheer, a Margi Avadhoot based in Bihar. He is suspected to have fled to Colombo.

The report has, however, not identified the end-user of the consignment. It is based on repeated interrogation of the An-26 crew by different agencies and gives a detailed account of how the armsdrop was planned.

The arrested Briton, Peter Bleach, has reportedly revealed less than he knows and is a key figure in the armsdrop conspiracy. His involvement is of greater import than he has projected.

While the absconding co-pilot, Kim Peter Davy, liaised with mercenaries on behalf of the recipient organisation during his Calcutta visit in September-October, Bleach was entrusted with the actual armsdrop operation.

Bleach, earlier identified as a former British defence person, is actually an ex-military intelligence man. After retiring, he became an arms dealer in North Yorkshire, Great Britain. Later, he extended his activities into South Africa and Zimbabwe. He disclosed his main client was the Bangladesh defence forces.

In September, two Danish nationals, Haestrup and Thuna, placed orders for 2,500 AK-47 and AK-56 assault rifles. When Davy got wind of this, he made discreet inquires.

At a meeting in Copenhagen later in September, the Danes introduced Davy to Bleach as the buyer of the weapons, beginning the Davy-Bleach collaboration. Davy, however, required a smaller quantity of weapons.

At the meeting, the three decided to drop the arms consignment in West Bengal. Bleach suggested it would be advantageous to use a cargo plane for the operation. Davy seconded the idea, saying similar jobs would come by easily in future.

At a meeting in Bangkok on September 29, Randy and Davy settled on Purulia as the arms delivery site and located it on an ordnance survey map. Bleach was present at the meeting.

Subsequently, it was decided the Purulia arms consignment would comprise 300 AK-47 and AK-56 rifles, 10 RPG/7, 25 automatic rifles, 100 hand grenades, two sniper rifles with night vision equipment, and adequate ammunition. Details of the air route were also worked out.

The next day, after returning from Bangkok, Bleach struck a deal with "Latavia", a bankrupt Latvian airline, for buying an An-26B aircraft. Latavia had been selling assets to tide over its financial crisis.

The deal was clinched at $ 250,000. Davy purchased it in November at Riga, the Latvian capital. Then he hired a seven-member crew from Latavia, including two ground engineers, a pilot, a co-pilot, a flight engineer and two assistants, for three months.

The aircraft could fly with the Latvian colours for a month or till it was registered in another country.

The crew members, except Bleach,

INVESTIGATION: THE ARMS DROP

are ethnic Russians. While six members are settled in Latvia, one is a Russian citizen. However, there were only five members on board. Two either opted out or their services were not required.

Subsequently, Davy acquired documents saying the An-26 had been purchased for Carol Air Services based in Hong Kong.

Apart from arranging the purchase of the aircraft, Bleach organised the arms consignment through Samuel Sieve, a London-based arms dealer of Polish origin with "extremely good contacts" in CENZEN, the Polish arms export corporation.

Sieve offered AK-47 and AK-56 rifles at $ 85.90 per piece. Owner of a company, Trade Investment, UK, Sieve then got in touch with his Bulgarian contact, Ivan Minkov.

Minkov delivered the consignment at Bourgas on December 10 in wooden crates marked "Technical Equipment."

Davy and his crew had earlier flown the aircraft there. They then loaded the consignment.

The report says the exact origin of the arms is not clear. But details given by the An-26 crew suggest they are of Bulgarian origin.

Arrangements for the delivery of the weapons cache at Bourgas is a pointer to the influence Bleach and Sieve have in Bulgaria. Minkov has access to all airports in that country.

The An-26 crew have not given a comprehensive account of the operation. Little is known about who ordered the weapons, who the end-users are and who provided funds.

Bleach's contacts with the Bangladesh military probably helped him obtain the end-user certificate, required for arms export. Crates with "Rajendrapur Cantonment" markings clearly suggest this.

There is possibly a nexus between Bleach and the Inter-Services Intelligence since passports of crew members, who stayed in Karachi for some time, do not bear Pakistani immigration stamps.

The fact that a Pakistani firm, Shaheen Enterprise, has links with Bleach and the Pakistan Air Force strengthens this possibility.

Davy top gold smuggler: CBI

BY CHANDAN NANDY

Calcutta, Dec. 31: Kim Peter Davy, who escaped from Mumbai's Santa Cruz airport after the An-26 aircraft was force-landed on December 22, is a top member of an international gold smuggling syndicate, according to findings by the Central Bureau of Investigation (CBI).

Davy has been in the limelight following revelations by the British pilot, Peter Bleach, and other Russian crew members involved in the Purulia arms drop.

CBI sources said the arms drop "is turning out to be a big thing with significant international ramifications."

Davy, a New Zealander settled in Hong Kong, is "essentially" a gold smuggler but is also involved in international arms and contraband smuggling.

Davy has visited Mumbai several times. In 1994, he visited the city four times. This year, he was in Mumbai, thrice,

Davy was earlier in touch with two "high profile" gold smugglers in Mumbai, a father-and-son team. They have shifted base to Dubai.

He also had links with two jewellery merchants and several hawala racketeers. He was associated with a lawyer, who was legal adviser to smugglers.

Davy's gold smuggling activi-

ties were financed by a Hong Kong-based Chinese, Mac, who allegedly supplied funds for the arms drop operation.

Davy has liaised with South Korean smugglers also. He has "unlimited cash reserves" and VISA credit cards issued by a Citibank branch in Hong Kong.

The CBI has traced Davy's links to Varanasi. He has made two four-day visits there.

Davy's links with Varanasi is significant because of its proximity to Nepal, widely used by smug-

glers to push contraband into India. Besides, the An-26 aircraft touched down at Varanasi airport for refuelling.

Davy may also have links with South East Asian narcotic syndicates. In this context, his frequent visits to Bangkok and Phuket, a coastal resort in Thailand, is significant.

Chinese-made sophisticated

INVESTIGATION: THE ARMS DROP

weapons are freely available in the Thai arms market. Narcotics are easily smuggled into the country because of the Golden Triangle.

In the 1980s, a Pakistani silver smuggling ring used Phuket as its base. It even owned a shipping company. But operations weakened in 1988-89 after Indian and Thai Customs authorities jointly seized several silver-consignments on the West Bengal, Tamil Nadu and Orissa coasts.

CBI sources have found Davy

tried to get the An-26 aircraft registered with Bangladesh Biman. He purchased the plane in November from Latavia, a bankrupt Latvian airline.

The agreement fell through because of a "financial hitch." Bangladesh Biman reportedly demanded too much money. The aircraft is not registered in Hong Kong either.

State Criminal Investigation Department and CBI sources said facts divulged by the An-26 crew strongly implicate the Ananda Margis.

Peter Bleach said Randy, identified as the liaison man for the recipient organisation and a suspected Margi, had a "weather-beaten face with long hair and beard and was barefeet." The CBI has already said Randy may have fled to Colombo.

■ CBI seeks Interpol help on crew, Page 4
■ End-users still to be identified, Page 8
■ CBI team leaves for Purulia, Page 8

CBI may probe Swiss bank link of armsdrop

BY CHANDAN NANDY AND
MEHER MURSHED

Calcutta, Jan. 5: With sketchy clues on the source of funds for the Purulia armsdrop, the CBI is planning to probe if the money used to purchase the Antonov-26 aircraft and the sophisticated weapons had been channeled into a Swiss bank account by the masterminds.

A CBI team is likely to leave for Geneva soon to inquire into the details of the monetary transactions through Swiss bank accounts. Separate investigations are being conducted to find out whether money was channeled through East European banks.

Investigations have revealed that Kim Peter Davy, the mastermind of the entire operation, had purchased the aircraft from Latavio, a bankrupt Latvian airlines, on wet lease by paying $250,000. The deal was struck in Riga, the Latvian capital.

Davy had got the An-26 aircraft registered in the Caicos Island, West Indies, a tax haven. The weapons were purchased from a Bulgarian company for about $165,000. CBI sources said each of the assault rifles was purchased for $ 85.

From the seized documents, including hotel bills, flight tickets, credit cards, share certificates and travellers cheques, and interrogation of the Briton, Peter Bleach, now under arrest and awaiting trial, have revealed that those who planned the operation had also used about $ 50,000 on "incidentals."

About $500,000 meant for the entire operation, has been detected. But the fact that cash payment was not made in any of the transactions has led the CBI to plan a strategy to extend its probe into some Swiss bank accounts.

Sources said the amount was transferred from a particular bank in Hong Kong, but the CBI was tightlipped about the Swiss bank where the money may have been deposited.

However, sources said such a huge amount must have been raised through smuggling of gold, silver and other contraband or by "forward trading." The CBI has found that Davy and his Indian contact, Deepak, alias Daya M. Anand, who is also absconding, belonged to an international gold smuggling racket operating from Hong Kong, Thailand, Singapore, Taiwan and West Asia.

"International mafia and smuggling gangs operating in arms and narcotics often transact business through Swiss bank accounts," an official said.

Sources said investigations are on to find out the details of a number of "secret" bank drafts and cheques which will confirm the name of the Swiss bank where they were deposited.

The financial transactions, the "grey areas" in the armsdrop probe, are being investigated through instrument of letters rogatory issued by the metropolitan magistrate's court here last year.

The CBI apprehends that investigations into possible transactions through a Swiss bank will be difficult because of tough banking laws in Switzerland.

The agency believes that "Mac," whose original name is Y. Hong Mac, and a wealthy Hong Kong-based "businessman" financed the whole operation. The CBI has apparently come to this conclusion because "Mac" was present in Riga when the deal to buy the aircraft was struck.

Sources said "Mac" was also present in an October 1995 meeting of the masterminds in Bangkok when the destination of the armsdrop was decided.

Armsdrop CIA scam: Bleach

FROM CHANDAN NANDY

New Delhi, Jan. 27: The British national, Peter James von Kalkstein Bleach, arrested in the December 1995 Purulia armsdrop case, has said the American Central Intelligence Agency (CIA) was behind the operation. "The armsdrop was a CIA scam," Bleach told the CBI.

Bleach, a former British intelligence official-turned-arms dealer settled in Yorkshire, says in his statement the armsdrop was a CIA-sponsored mission. However, this does not form part of the huge corpus of documents already submitted before the City Civil and Sessions Court in Calcutta, where Bleach, the five Latvian crew members and the brother of a key accused went on trial today.

Bleach was arrested along with five Latvian crew members of the Antonov-26 aircraft, which was used to paradrop the arms consignment on December 22, after the plane was made to land at Mumbai. The mastermind, Kim Peter Davy, identified as a Danish national, Niels Christian Nielsen, escaped under mysterious circumstances from the airport.

Bleach's disclosure that the armsdrop was an American intelligence operation was based on his belief that during the same time as the armsdrop was planned, the CIA had undertaken a number of clandestine operations in Western Europe.

Bleach was candid while providing details to the CBI on most aspects of the armsdrop: the planning stage in Copenhagen; the main conspirators; the company from which the arms were purchased in Bulgaria and the destination of the consignment; the route to be taken by the AN-26; a part of the financial transactions;

rough identities of Acharya Suranjanananda Avadhut, alias Satyendra Narayan Gowda, alias Randy, and Deepak, alias Daya M. Anand.

In fact, Bleach had even quoted Davy as saying the arms and ammunition were meant for "people oppressed by Communists in West Bengal", though he had no clue at that time that the cache of weapons was actually meant for the Ananda Margis.

What Bleach did not reveal, however, were his "contacts" in the Bangladesh Armed Forces, especially Major-General Subed Ali Bhuyan, Principal Staff Officer in the Armed Forces Division during Begum Khaleda Zia's tenure, who actually signed the end-user certificate for the consignment.

Bleach passed all information relating to the planning of the armsdrop and a rough idea of where the weapons would be air-dropped to British intelligence. And there is evidence to prove he was keeping British intelligence and Scotland Yard updated on all developments.

But he says: "From the beginning I thought that the entire project had an American feel. A good overall plan has been spoilt by really bad attention to detail — a problem which has been a feature of US-sponsored operations for many years."

Bleach should know. He had served the British intelligence as an undercover agent against the Irish Republican Army. Pinpointing where the planners went wrong, Bleach says: "While they seem to have been careful to provide me with misinformation on a variety of important points, it is amazing that they actually allowed me to know the end destination. I should never have been allowed to know it."

ACKNOWLEDGEMENTS

This book could not have been conceived without the deep insights and information that dozens of friends and sources in the Indian security establishment and abroad provided me over several months. Some have preferred to remain, as they did while in service, anonymous. I respect their decision and cannot thank them enough for the hours and hours of fruitful discussions and brainstorming over this most mystifying case. Many of them invited me to their homes, others I met in restaurants and cafés across cities in India. I thank them for their warmth and generosity.

No words are enough to express my gratitude for my sources and friends in India's intelligence agencies, and investigators in the CBI—very few in number—who chose to be identified and spoke freely, offering valuable information and tips. I thank, especially, a former director of the IB who unhesitatingly reviewed sections of the draft typescript and offered critical comments on them. He was in many ways a guide and an incisive critic rolled into one. A similar role was played by a former deputy chief of the R&AW, a passionate and an unremitting critic of his own organization and the Indian security system. I thank him profusely for providing regular advice and suggestions, besides sharing and discussing important and sensitive elements of the case, not to speak of the nuances of covert intelligence operations.

There were other, very senior retired intelligence officers who stood by their bounden duty of keeping the country's secrets and politely refused to speak or engage me in conversation. I respect their decision and thank them, for their stand pushed me on to relentlessly pursue the story.

A friend, Indian Air Force Wing Commander (retired) K.T.

Sebastian, generously analysed some aviation-related technical data. I thank him for being patient with all sorts of questions that I constantly put across to him, about matters of which I had little knowledge. My thanks to another IAF officer, a retired air vice-marshal, who gladly shared details of a few secret operations in India's neighbourhood. I am indebted to Brian Wood at Amnesty International, London, for sharing with me crucial international legal instruments and rightly cautioning me to be 'careful' about how certain players in the global black market trade in SALWs have been portrayed by international institutions and Western governments.

I also thank Peter Bleach, one of the important dramatis personae of this bewildering case, for speaking to me about his angst and for voicing his questions, doubts and suspicions. I sympathize with him and his friends in England for his most unfortunate years of incarceration at Calcutta's Presidency Jail.

Thanks to Vir Sanghvi for the lucid foreword. My editor at *Hindustan Times*, Vir has a phenomenal memory of some of the intricate details of the Purulia arms drop. To Subir Bhaumik, a fellow 'Purulia investigator' and a widely acknowledged expert on India's Northeast and insurgency, many thanks for his generous comments and endorsement.

I admire the manner in which the committed editorial team at Rupa Publications, especially Elina Majumdar and Amrita Mukerji, speedily edited the typescript.

I thank a select bunch of friends, including a former colleague in *The Telegraph*, Meher Murshed, who were a constant source of encouragement for this book project. I appreciate the prompt affirmative action taken by *The Telegraph* editor-in-chief, Aveek Sarkar, and librarian, Shakti Roy, on my application seeking reproduction in this book of some of my stories that appeared in the newspaper between 1995 and 1999.

Finally, I thank my mother, Krishna Nandy, for being a patient listener to every advance I made with the story and for gently pushing me into completing this book.